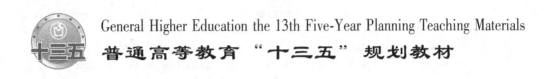

普通高等教育"十三五"规划教材

Modern Technology of Mining Engineering
采矿新技术

何富连　陈见行　张守宝　编著

Beijing

Metallurgical Industry Press

2023

Abstract

This textbook systematically introduces domestic and overseas new mining technologies. The contents include: new mine exploiting technologies in domestic and overseas countries, research progress on the mining pressure and its controlling in stopes and entries, deep mining and research progress on dynamic pressure, scientific exploiting system in mining, exploiting key technologies in new regional mining. In the textbook, the domestic and overseas advanced research achievements and in-situ practical experiences in mining engineering are comprehensively reflected.

This textbook can not only be regarded as the education reference book for senior undergraduate andpostgraduate students related to mining engineering, safety technology and engineering, geotechnical engineering in colleges and universities, but also for teachers, researchers, engineers, designers and science management staff who are working in mining engineering to read.

图书在版编目（CIP）数据

采矿新技术＝Modern Technology of Mining Engineering：英文/何富连，陈见行，张守宝编著．—北京：冶金工业出版社，2020.3（2023.1重印）
普通高等教育"十三五"规划教材
ISBN 978-7-5024-8395-1

Ⅰ.①采… Ⅱ.①何… ②陈… ③张… Ⅲ.①矿山开采—高等学校—教材—英文 Ⅳ.①TD8

中国版本图书馆 CIP 数据核字（2020）第 021770 号

Modern Technology of Mining Engineering

出版发行	冶金工业出版社	电　　话	（010）64027926
地　　址	北京市东城区嵩祝院北巷39号	邮　　编	100009
网　　址	www.mip1953.com	电子信箱	service@mip1953.com

责任编辑　李培禄　美术编辑　吕欣童　版式设计　孙跃红
责任校对　郑　娟　责任印制　窦　唯

北京虎彩文化传播有限公司印刷
2020年3月第1版，2023年1月第2次印刷
787mm×1092mm　1/16；16印张；390千字；246页
定价45.00元

投稿电话　（010）64027932　投稿信箱　tougao@cnmip.com.cn
营销中心电话　（010）64044283
冶金工业出版社天猫旗舰店　yjgycbs.tmall.com
（本书如有印装质量问题，本社营销中心负责退换）

Preface

In recent years, there has been a tremendous progress on the coal mining and production technology in China, which effectively promotes the progress of coal industry. The coal industry in China is continuously moving towards to internationalisation and globalisation. Therefore, colleges and universities related to mining should grasp the pulse of times tightly, cultivating more intelligent international talents for the society. Also, undergraduate students within the mining engineering major should grasp not only professional fundamental knowledge but also professional English. In this way can students better understand the main mining methods, technologies and equipment used in countries all over the world and grasp the hot spots in the discipline of mining engineering, becoming brilliant professional talents who have international perspective.

To fulfil the requirement of education, research and in-situ production related to mining and provide education reference books for students related to mining to understand the cutting-edge technologies in the discipline of mining engineering, the authors wrote and compiled this textbook after referring to a large number of domestic and overseas original documents related to new mining technologies and the successful experiences regarding coal mine production, based on summarising the results in years of research and education.

To expand students' knowledge, this textbook regards the coal mining technology as the main part. Also, it introduces new technologies in preventing mine disasters and in exploiting other resources. These include mine exploiting technologies in domestic and overseas countries, research progress on the mining pressure and its controlling in stopes and entries, deep mining and research progress on dynamic pressure, scientific exploiting system in mining, exploiting key technologies in new regional mining.

This textbook totally includes 7 units and 28 special technical contents. Among them, typical related engineering cases are included in the main contents. The authors expect that after readers learn this course, they can fully understand the development trend related to domestic and overseas new mining technologies. Furthermore, the authors expect that this textbook could use a little to get a big, or the readers can refer and grasp more new mining technologies based on their own actual situation, learning to meet practical needs indeed.

The publication of this textbook was financially supported by the National Natural Science Foundation of China "Main Roof Breaking Mechanism and Reverse Increasing Underground Pressure Control of Gob-side Entry with Longwall Top Coal Caving Mining in Extra-thick Coal Seam" (Grant Number: 51974317), the Yue Qi Distinguished Scholar Project (Grant number: 800015Z1138), China University of Mining and Technology (Beijing) and the Fundamental Research Funds for the Central Universities" (Grant number: 800015J6). This textbook also received the support from the School of Energy and Mining Engineering, at the China University of Mining and Technology (Beijing) and the help from members in the research team of the authors. Collection of typical cases in this textbook received the care and guidance from collaborated coal companies and the staff. Then, faithful thanks are given to all units and persons that provided support to the publication of this textbook.

During the writing and compiling process, a large number of domestic and overseas references were cited and referred. Loyal thanks are given to all authors of those documents again. Due to the authors' limited knowledge, there might be some mistakes and flaws in this textbook. Pease do not hesitate to correct us.

The authors
March 2020

Contents

Introduction 1

1 New technologies of mine exploitation in China 10
   1.1 Top coal caving mining technology in extremely large-scale mines 10
      1.1.1 Geological production conditions of the Tashan Coal Mine 10
      1.1.2 Coal mining techniques 10
      1.1.3 Equipment configuration 11
      1.1.4 Roof management 13
      1.1.5 Ground pressure observation 13
      1.1.6 The high production and high efficiency mode of Tashan Coal Mine ... 14
   1.2 Large mining height technology in thick coal seams 15
      1.2.1 Geological production conditions of the Shangwan Coal Mine 16
      1.2.2 Optimisation of the production system in Panel 4# in coal seam 12# .. 16
      1.2.3 Typical example of the set of fully mechanised working face equipment for large mining height (8.8m) 19
      1.2.4 Innovative outcomes 21
   1.3 The fully mechanised coal exploiting technology in thin coal seams 22
      1.3.1 Geological production condition of the Laoshidan Coal Mine 22
      1.3.2 Technical design and equipment selection of the fully mechanised mining in the thin coal seam 23
      1.3.3 Reinforcement technology of the horizontal roadway in the thin coal seam ... 27
      1.3.4 Ground pressure observation 28
   1.4 Mining and safety technologies of the coal seams with extremely adjacent distance 28
      1.4.1 Geological production conditions of the Wuhushan Coal Mine 28
      1.4.2 Determination and its controlling of the proper position of the mining roadways and the open-off cut 30
      1.4.3 Designing selection of the fully mechanised equipment 34
      1.4.4 Ground pressure observation and gas safety management in exploiting of the extremely adjacent coal seams 35

2 New mining technologies in the coal industry in overseas countries 40
   2.1 Advances in mining and safety technologies in USA 40

2.1.1　Safety technological developments ……………………………………… 40
2.1.2　Mine safety legislative developments & safety organizations in the U.S. ……… 45
2.2　Borehole mining ………………………………………………………… 46
2.2.1　How BHM works ………………………………………………… 46
2.2.2　Current BHM applications ……………………………………… 48
2.2.3　New BHM applications ………………………………………… 49
2.2.4　When to use BHM ……………………………………………… 50
2.3　The mining technology in open-pit mining …………………………… 51
2.3.1　Classification of the open-pit mining methods ………………… 51
2.3.2　Technical procedures in the open-pit mining method ………… 52
2.3.3　New technologies and new equipment in overseas open-pit mining …… 52
2.3.4　Development tendency of the open-pit coal mining …………… 55

3　Research progress on the ground pressure in the longwall face and its controlling technology ………………………………………… 58

3.1　Forecasting and the research progress of the mine roof disasters ……… 58
3.1.1　Mine roof disaster state and the rock mass environment ……… 58
3.1.2　Classification of the leaded disaster and the forecasting theory of the mine roof disasters ………………………………………………………… 60
3.1.3　Index system and principle of the mine roof disaster forecasting …… 66
3.1.4　Demonstration engineering of the project application ………… 69
3.2　Study of the ground pressure law and the support rationalisation of the extremely large-scale fully mechanised top coal caving working face ……… 71
3.2.1　Geological production condition of the Anjialing underground coal mine 2# …… 71
3.2.2　Measuring of the floor pressure ratio and the adaptability of the hydraulic support basement ……………………………………………………… 72
3.2.3　Relationship between the type of the fully mechanised top coal caving hydraulic support and the supporting performance and the rationality of the hydraulic support type ………………………………………………………… 75
3.2.4　Design of main parameters of the fully mechanised caving shield hydraulic supports with two columns to support the roof ……………………… 77
3.3　Controlling of the hydraulic support-surrounding rock mass in the fully mechanised top coal caving working face with large dip angle under the complicated condition …………………………………………………… 80
3.3.1　Engineering profile of the fully mechanised caving working face with large dip angle under the complicated condition ……………………………… 80
3.3.2　The hydraulic support collapsing accident with large dip angle and its controlling parameters ……………………………………………………… 82
3.3.3　Leakage malfunction detecting technology of the hydraulic supports …… 87

3.3.4 Safety mining controlling practices in the fully mechanised caving working face with large dip angle in the Ge'ting Coal Mine ······ 88

4 Research progress on the ground pressure around the roadway and its controlling technology ······ 92

4.1 Rock bolt reinforcement and the roadway surrounding rock mass reinforcing ······ 92
4.1.1 Principles for rock support ······ 92
4.1.2 Bolt-rock interaction ······ 94
4.1.3 A practical problem of rebar bolts ······ 97
4.1.4 Examples for rock support design ······ 98

4.2 Circular cooperative controlling of the extremely large cross-section roadways with water spraying and water burst in the area of the coal mass and mudstone chamber group ······ 99
4.2.1 Geological production condition of the Xinzhi Coal Mine ······ 99
4.2.2 The failure mechanism and controlling principle of the roadways with extremely large cross-section in the chamber group ······ 101
4.2.3 Circular cooperative controlling technology in terms of grouting, anchoring, supporting and casting ······ 101
4.2.4 The circular cooperative controlling technical scheme in terms of grouting, anchoring, supporting and casting ······ 104
4.2.5 In-situ practice and observation ······ 107

4.3 The cable truss combined reinforcement in the open-off cut with the composite mudstone roof and large mining height ······ 108
4.3.1 Geological production condition of the Bailong Coal Mine ······ 108
4.3.2 Rock bolt and cable bolt problems in the mudstone coal roadways with large span ······ 109
4.3.3 The high pretension cable bolt truss controlling system ······ 110
4.3.4 The composite reinforcement scheme of the cable truss system in the open-off cut ······ 113

4.4 The asymmetric failure mechanism and controlling of the surrounding rock masses in the gob-side tunnelling in the fully mechanised caving area with large cross-section and intense mining ······ 116
4.4.1 The geological production condition of the Wangjialing Coal Mine ······ 116
4.4.2 The controlling problem of the surrounding rock masses in the fully mechanised caving coal roadway with large cross-section and intense mining ······ 117
4.4.3 The asymmetric failure mechanism of the coal and rock masses in the roof of the gob-side coal roadway in the fully mechanised caving ······ 118
4.4.4 The controlling system of the newly developed high pretension

cable truss system ··· 121
4.4.5 The in-situ reinforcement industrial experiment on the fully mechanised caving coal roadway with large cross-section and intense mining ··· 124
4.5 The gob-side roadway retaining technology and the technical optimisation and improvement in the section of the fully mechanised mining ··· 127
  4.5.1 The surrounding rock mass controlling principle of the gob-side roadway retaining ··· 127
  4.5.2 The controlling strategy of the surrounding rock masses in the gob-side roadway retaining ··· 128
  4.5.3 The technique of the gob-side roadway retaining system ··· 132
  4.5.4 In-situ tests of the gob-side roadway retaining ··· 134
4.6 Engineering design and practice of the pressure relieving in the high stress and weak coal roadway ··· 136
  4.6.1 Geological and productional conditions ··· 136
  4.6.2 Simulation and design of the destressing scheme ··· 137
  4.6.3 Destressing practice ··· 140

5 The challenging problems of deep mining and dynamic pressure study ··· 142

5.1 The challenging problem of surrounding rock mass controlling in the deep mining and the engineering technology ··· 142
  5.1.1 The reinforcement technology of the roadways in the deep area in China and overseas countries ··· 142
  5.1.2 The development of the deep roadway reinforcement theory ··· 144
  5.1.3 The deep roadway reinforcement technology and typical example ··· 146
5.2 The challenging problem and the solving practices of the floor heave in the weak rock roadways in the deep mine ··· 149
  5.2.1 Mechanism of the roadway floor heave ··· 150
  5.2.2 The controlling theory and technology of the floor heave of the roadways ··· 155
  5.2.3 Materials and equipment used in controlling the roadway floor heave ··· 157
  5.2.4 Engineering cases of processing the deep mine floor heave ··· 159
5.3 The bump and outburst hazard coupling and its prevention technology of the island fully mechanised working face with large mining height and high risks ··· 162
  5.3.1 The island working face bump and outburst coupled parameter studying and analysing ··· 162
  5.3.2 The coupled mechanical model and occurring mechanism of the rock burst and outburst ··· 163
  5.3.3 Numerical simulation of the stress field in the island section and the proper

        position of the horizontal roadway ··· 164
   5.3.4   Experimental study on the bump tendency of the coal and rock masses in the
        is land working face ·· 166
   5.3.5   The bump hazard grading and the partition predicting of the mining zone with the
        large mining height ·· 166
   5.3.6   The prevention scheme and parameters of the bump and outburst ············· 169
5.4   The disaster mechanism and adjusting of the rock burst and underground
    reservoir ·· 172
   5.4.1   The variation law and mechanism of the rock bursts with different types ······ 172
   5.4.2   Hazard estimation and predicting of the rock burst ································ 174
   5.4.3   The dynamic adjusting and controlling method of the rock burst developing
        process ·· 177
   5.4.4   Practices on the deep buried diversion tunnel and the water discharging tunnel in
        the Jinping Level two hydropower station ··· 177
   5.4.5   The disaster mechanism and the prevention theory of the underground reservoir
        group ·· 179

6 Scientific exploiting system of the coal industry ································ 183
6.1   New development on the backfill coal mining technology and the paste
    backfill mining technology ··· 183
   6.1.1   The rock strata movement controlling theory with the backfill coal mining ······ 183
   6.1.2   The integrated technology with the backfill and coal miningcombined [19] ······ 184
   6.1.3   The technology and methods of recycling the strip coal pillar with paste backfill
        mining [20] ·· 187
6.2   The coal resource exploiting with the replacement of the waste gangues
    in the fold and fault area in which the high abutment pressure occurs ······ 188
   6.2.1   Geological production condition of the Xinsan Coal Mine ························ 188
   6.2.2   Support requirement and controlling scheme for the coal roadways with large
        cross-section in the high stress domain ·· 190
   6.2.3   The developing and techniques of the exploiting equipment for the coal resources
        with the waste gangue replacement ··· 191
6.3   The combined mining technology of coal and gases, and the underground
    gasification ··· 196
   6.3.1   Combined exploiting of coal and gases ·· 196
   6.3.2   The underground gasification ··· 200
6.4   The exploiting of the mine without drainage and the underground water
    storage technology ·· 201
   6.4.1   The underground reservoir construction technology of coal mines ············· 202
   6.4.2   The safe operation technology of the underground reservoir in coal mines ······ 206

6.4.3 Application effect 209
6.5 Intelligent mining technology 210
  6.5.1 The electro-hydraulic remote coordinated controlling system with groups 211
  6.5.2 Reliability of the set of the equipment system in the fully mechanised mining 212
  6.5.3 Self-moving hydraulic supports following the shearer and the intelligent coal caving technology 214
  6.5.4 Digital mine management platform system 216
  6.5.5 Application effect 218

7 The core technology in exploiting the mining resources in new zones 219
7.1 The exploration and exploiting system of the ocean ore resources 219
  7.1.1 Flammable ice 219
  7.1.2 Manganese nodule 223
  7.1.3 Petroleum 226
7.2 The distribution and the exploiting expectation of the mineral resources in the polar region 228
  7.2.1 The mineral resources in the polar region is various and abundant 228
  7.2.2 Exploiting the resources in the polar region 232
  7.2.3 Exploiting expectation of the resources in the polar region 236
7.3 Exploiting characters of the highland mining industry, and the exploiting technology and equipment 237
  7.3.1 The mineral resources in the highland and the Alpine region 237
  7.3.2 The technic technologies of the open-pit mining in the highland mining 238

References 245

Introduction

Mining is a complicated production process. Technologies including geology, surveying, tunnelling, reinforcement, coal extraction, transport, lifting, ventilation, drainage, electronic power supply, safety, mechanisation and automation may be synthetically applied. Meantime, advanced business administration and operation methods should be carried out. The mining industry proposes requirement on the technology and equipment. Meanwhile, it rapidly develops because of the improvement of those technologies and equipment. The mining technology includes multiple contents of branches such as shaft sinking and drifting, mining methods, ventilation safety technology, mining machinery, power supply in mining, mining economics and management, and mining geology.

I Mining industry requires to develop new mining technologies

The China Geological Survey compiled and published China Minerals Resources 2018, which showed that until the end of 2017, the remaining reserves of coal were 1666.67 billion tons. Coal accounted for 76% and 66% in the primary energy production and consumption structure in China. In Development Planning Sketch of Energy in Medium and Long term (2004 ~ 2020), the government determined that China would insist on the energy strategy that coal was regarded as the main body. Power was regarded as the centre. Oil, gas and new energy developed roundly. The report of State Energy Development Strategy from 2030 to 2050 provided by China Academy of Engineering proposed that the coal production in 2050 should be controlled at approximately 3 billion tons. Coal will be the principle energy for a long term in China.

China has a vast territory. The topographic and geological conditions are pretty complicated. Therefore, China has a high-level requirement on the mining technology, which must continuously reform with the era development. In this way, the efficient, safe and clean exploitation and utilisation of coal resources can be realised, promoting the healthy and sustainable development of coal industry. The significance of developing new mining technology is reflected in the following aspects:

(1) Utilising new mining technology can promote economic benefits.

To promote the economic benefits of companies, measures that can decrease production costs can be selected. Furthermore, convenient and efficient production situation can be utilised to promote economic benefits. Especially in the coal mining industry, machinery is used to replace labour in most links. Because of the continuous development on the number of new machines and new technologies, the production efficiency in coal mines also increases continuously. If companies can utilise new technologies to exploit coal mines times, this will not only promote the coal mine production but also decrease labour work. In this way, the labour cost can be reduced. Furthermore, this can prevent safety disasters, apparently increasing the safety of coal mine exploitation. Utilising new technologies to carry out production in companies can not only fulfil the requirement of scientific development concept but also display the value of advanced scientific achievements.

(2) Relied on new mining technologies, companies can not only make the coal enterprise management more normative but also increase competitiveness.

Companies will abandon those obsolete and behindhand mining facilities. Then, new mining technologies will be used, which can make the enterprise management more normative. When coal mine companies use new mining technologies, they should rely on associated new equipment, which can guarantee the whole quality and safety performance. In this way, the management of company mechanical equipment can become large-scale and unified. Furthermore, this make it easy to examine and maintain the mechanical equipment. In addition to that, compared with other competition staff, if companies can use new mining technologies in advance, the gap between them can be enlarged. In the production and construction process, more prominent superiority can appear, therefore acquiring more economic benefits.

(3) That coal mine companies rely on new mining technologies can stipulate the coal production system, decreasing the possibility of disasters.

Through conducting research on the coal mine production in China, it was found that there were still certain issues. In fact, the distribution of Chinese coal mines is scattered and there is certain difference on the scale. Furthermore, the economic level of the corresponding areas is different. Therefore, apparent difference occurs on the mining technologies, which is easy to induce safety accidents. In the current stage, after the government nationalises several coal mines, the number of accidents apparently decreases. This reveals that uniform standard management can effectively prevent safety accidents. Coal mine companies should cognise that selecting new mining technologies to conduct mining. Furthermore, management system should be completed and advanced technologies should be used to conduct mining activities, which can reduce not only cost but also waste. Through setting normative production system in the coal exploitation industry in China, transferring can be fulfilled timely when the state issues tasks to the coal mine companies or unifies indexes. This prevents other potential safety hazards which may be induced by issues related to technologies and equipment management. This makes the state coal mine exploitation industry achieve more apparent development.

II Principal coal bases in China

At the end of 2016, the state issued the State Mineral Resources Planning from 2016 to 2020, proposing that 14 coal bases including Shendong, Northern Shanxi, Middle Shanxi and Western Shanxi will be emphatically constructed. Furthermore, 162 state projected coal mines were designated.

(1) Shendong Base: The Shendong Base includes Shendong, Wanli, Zhunge'er, Baotou,

Wuhai and Fugu mines. Its area is approximately 12.6km². The coal resource reserves with the buried depth lower than 1000m are 319.015 billion tons.

(2) Jinbei Base: The Jinbei Base includes Datong, Pingshuo, Southern Pingshuo, Hebaopian and Lan County mines, accounting for over 1/3 of the total area of the Shanxi Province. The total area with coal is 6700km². It is important coal and power energy base in China. The resource reserves are 111.852 billion tons.

(3) Jinzhong Base: The Jinzhong Base includes Xishan, Dongshan, Fenxi, Huozhou, Liliu, Xiangning, Huodong and Shixi mines. Its area is 27475.06km², in which the area with coal deposits is 23861.39km². The resource reserves are 207.273 billion tons.

(4) Jindong Base: The Eastern Shanxi Base includes Jincheng, Lu'an, Yangquan, Wuxia mines. Its area is 15060km². The resource reserves with the buried depth deeper than 1300m are 122.721 billion tons.

(5) Shanbei Base: The Shanbei Base includes 4 mines, namely Yushen, Yuheng, Wubao and Zichang. Its area is 25950.82km². The resource reserves are 200.019 billion tons.

(6) Huanglong (Huating) Base: The Hunaglong (Huating) Base includes Binchang (including Yonglong), Huangling, Xunyao, Tongchuan, Pubai, Chenghe, Hancheng, Huating mines. Its area is 18877.63km², in which the area with coal deposits is 11084km². The resource reserves are 48.16 billion tons.

(7) Luxi Base: The Luxi Base includes Yanzhou, Jining, Xinwen, Zaoteng, Longkou, Zibo, Feicheng, Juye, Northern Yellow River mines. Its area is 55962.82km² and the resource reserves are 34.8 billion tons.

(8) Lianghuai Base: The Lianghuai Base includes the Huaibei mine and the Huainan mine. Its area is 19000km², in which the area with coal deposits is 9651km². The resource reserves are 51.68 billion tons, in which the Huainan mine (with the buried depth deeper than 1500m) has 39.98 billion tons and the Huaibei mine (with the buried depth deeper than 1200m) has 11.7 billion tons.

(9) Henan Base: The Henan Base includes Hebi, Jiaozuo, Yima, Zhengzhou, Pingdingshan, Youngxia mines. Its area with coal deposits is 10849km² and the resource reserves are 56.011 billion tons.

(10) Yungui Base: The Yungui Base includes Panzhou, Puxing, Shuicheng, Liuzhi, Zhina, Northern Guizhou, Laochang, Xiaolongtan, Zhaotong, Zhenxiong, Enhong, Junlian, Guxu mines. Its area is 0.469 million km², in which the planned exploitation area is 4234km². Except that the Xiaolongtan Mine locates in the Kaiyuan City which is in the south of the Kunming City, other mines are located in the junction area of three provinces, namely Sichuan, Guizhou and Yunnan. The resource reserves are 95.348 billion tons.

(11) Mengdong (Dongbei) Base: The Mengdong (Dongbei) Base includes Zhalainuoer, Baorixile, Yimin, Dayan, Huolinhe, Pingzhuang, Baiyinhua, Shengli, Fuxin, Tiefa, Shenyang, Fushun, Jixi, Qitaihe, Shuangyashan and Hegang mines. Its area is 27264.7km² with the resource reserves of 114.917 billion tons. Among them, the Eastern Mongolia area has 85.787

billion tons and the Liaoning Province has 7.44 billion tons. The Heilongjiang Province has 21.69 billion tons.

(12) Jinzhong Base: The Jinzhong Base includes large-scale coal basins such as Fengfeng, Handan, Xingtai, Jingxing, Kailuan, Yu County, Xiahuayuan in Xuanhua, Northern Zhangjiakou and Pingyuan. Its resource reserves are 44.634 billion tons, in which the resource reserves with the buried depth lower than 1000m accounts for 42%. In addition to that, the resource reserves with the buried depth ranging from 1000m to 1200m account for 31% while the resource reserves with the buried depth ranging from 1200m to 1500m account for 27%.

(13) Ningdong Base: The Ningdong Base includes Shizuishan, Shitanjing, Lingwu, Yuanyanghu, Hengcheng, Weizhou, Majiatan, Jijiajing, Shizheyi and Mengcheng mines. Its area is approximately 18876km^2, in which the area with coal deposits is approximately 2105km^2. The resource reserves are 29.399 billion tons.

(14) Xinjiang Base: Coal resources in Xinjiang mainly distributes in three basins, namely Zhunge'er, Tulufan-Hami and Yili. Its resource reserves are 280.435 billion tons. Principal coal basins include Zhundong Coal Basin, Zhunnan Coal Basin, Tuoluogai Coal Basin, Tuha Coal Basin, Yili Coal Basin, Kubai Coal Basin and Yanqi Coal Basin. Under the background that the quantity of resource reserves in Chinese mid-east coal mines decreases sharply, Xinjiang has already become important energy succeeding area and strategic energy reserving area in China. According to the state's planning of further development regarding Xinjiang, the annual coal production of Xinjiang will account for 20% of the whole annual production of the country in the future.

III Current situation and challenging in coal mine exploitation

China is a big country in coal production. Currently, the coal industry in China has already grasped the ability of designing, construction, equipping and managing open-pit coal mines with ten million tons and medium-scale and large-scale mines. Modern mechanised equipment, mechanised excavation equipment and large-scaled peeling, mining, transporting, reinforcing equipment for open-pit mines are widely used in medium-scale and large-scale coal mines. However, due to unbalanced development, it is overall in the relatively advanced level. The main reasons leaded to this situation are the coal industry basement is weak and the quality of employees is uneven. Furthermore, the equipment reliability needs further improvement and the promotion progress of new techniques, technologies and equipment is slow.

With the requirement of coal industry sustainable development on resources, safety and environment continuously improves, mining technologies are also confronting a series of new issues and challenge [1].

(1) The fundamental theory research on coal resource exploitation relatively lags behind, needing to be solved urgently. First, the condition of coal resource reserves is complicated and there are many natural disasters. Massive significant key technologies need to be solved. Secondly, coal production increases. The harmonious development between large-scale exploitation and ecological environment urgently requires that the scientific problem of coal green mining and

clean utilisation.

(2) The guaranteeing ability of coal resources is relatively low. The emphasis of coal exploitation gradually moves westward. The coal resource reserves in China are 1666.673 billion tons, in which the fundamental reserves are 317.62 billion tons. The coal resources in western areas are more than 60% of the whole quantity. The resources in eastern area of China are going to be exhausted. The mining technology in deep section needs to be solved urgently. The exploitation strength in the middle enrichment section of China is high. However, the ecological environment is weak. The resource exploitation and the technology of coordinating environment should be studied. The western area is rich in resources. The emphasis of coal exploitation gradually moves westward. However, high strength exploitation technologies that can adapt to the characteristics of strata in the western area are deficient. Technologies including coal conversion, transferring, exploitation of extremely thick coal seams and fire prevention on coal seams that may easily self-ignite need to be solved urgently.

(3) Adjustment of coal industry structure and promoting the coal mine modern level propose new requirement on the innovation of coal technology. The ability of mechanised mining and mechanised excavation, the reliability of key components, artificial controlling, coal mine management informatisation and the modern level of safety monitoring need to be improved. Safe and efficient mining technologies of coal seams that are difficult to be exploited, such as extremely thin, extremely inclined, apparent variation of reserves and seriously threatened by disasters like extruding of coal and gases, need to be solved urgently. Meantime, the mechanised exploitation technologies in small scale coal mines and small block coal resources need to be solved urgently.

(4) Intensive development, clean development, safe development and sustainable development propose higher requirement on the innovation of coal technology. The fully delicate survey technology and equipment of coal resources in China relatively lag behind. The geological guaranteeing level needs to be improved. Aspects including water-preserved mining, management of subsidence induced by coal mining, management of dusts and noises, prevention of heat damage and utilisation of geothermal energy, and utilisation of underground water should be resolved urgently. A series of significant key equipment and technologies require to be studied, such as exploitation of coal seam gases, formation mechanism of coal mine gas disasters and early warning of monitoring and prevention, underground backfilling, coal mining under water, roadways and buildings, especially the threats coming from high pressure water and rock burst prediction together with prevention.

IV Development tendency of mining technologies in China

The life cycle of mine exploitation includes exploration, exploitation, beneficiation and reclamation. The exploitation links of mineral resources are related to utilisation of multiple disciplines and collaborative innovation. With the rapid development of multiple disciplines, a number of new technologies that can promote the development of mining occur, similar to bamboo shoots after a spring rain. New concepts such as intelligence, informalisation, greening mining are put forward. Overall, the mining industry in China is going to display the following new tendencies [2].

(1) Deep mining: With the shallow resources to be continuously consumed, an increasing number of domestic and international mine sites entered deep mining stage. Based on the geological condition, mining technology level, mining equipment level and mining pressure behaviour around roadways, it is generally regarded that the mining depth which is equal to or deeper than 800m is belonged to deep mining. However, currently, the mining depth in South Africa is over 3000m. New mining depth induces new challenges for mining staff. For example, the status of "three high indexes with one disturbance" (high in-situ stress, high geothermal temperature, high karst water pressure and one mining disturbance) in deep surrounding rock masses leads to appearance of mine disasters (rock burst, large deformation of surrounding rock masses, water burst, high karst water pressure and high mining disturbance).

(2) Intelligent mining: Intelligent mining focused on developing and exploiting mining equipment which regard intelligence as the principal part. It relies on information collection and rapid communication, accurate orientation and intelligent navigation, intelligent mining blasting, intelligent controlling during the production process, information-based platform in decision-making and management. This promotes the utilisation rate of mineral resources and economic benefits of mining companies.

(3) Green exploitation: During the period of "13th Five-Year Plan", due to the variation of macroscopic economic situation, mine sites should develop along the direction of no waste and green mining. In the time of promoting production efficiency, coordinating the environment protection should be realised, fulfilling the sustainable development of mining companies. Green exploitation means realising and treating all kinds of resources in the mine site area from the perspective of generalised resources. Its basic springboard is prevention and relieving the harmful effect of mining on environment and other resources as far as possible, then acquiring the best economic, social and environmental benefits. Green exploitation is a mine exploitation mode that fully considers the environment influence and resource consumption. It is a component of sustainable development. Under this aspect, mining is conducted without wastes. The waste rocks will not be put outside of the pit. Tailings will be backfilled directly to the gob area. No ground tailing warehouse will be constructed. The ground surface will not be destroyed. Therefore, green mining will be realised.

(4) Exploiting in the deep sea and highland: Currently, in the world, the human beings are confronting with the increasing worsening of land resources and energy. In the deep sea and highlands, there are abundant kinds of metal and non-metal mine resources, energies and biological resources. The exploiting potential is high. However, the utilisation extent of resources is low. Therefore, in the future, exploiting in the deep sea and highlands has pretty bright prospect. Improving the exploiting and utilisation of the resources in the deep sea and highlands has important significance for realising the balanced development between the economic increasing and the integrity of the environment.

(5) Ecology reconstruction in mine sites: The ecology reconstruction in mine sites is the key or hinge of district sustainable development. Therefore, exploitation and utilisation of mine resources

disturb the ecological environment system for specific areas maximally. The destroying strength is also maximal. Reconstructing the land production ability in the mine area and maintaining the health of ecological system have significant practical role on agricultural production and environment protection. Managing the environmental pollution and geological disasters in mine sites involves ecological economics, environment monitoring technology, water pollution controlling engineering, environmental chemistry, disaster prevention engineering, land reconstruction and improvement technology, ecological protection technology and vegetation restoration technology.

(6) Coordinated exploiting of coal and associated resources: The coordinated exploiting of coal and associated resources regards the ecological protection as the premise and regards the accurate exploration and intelligent mining as the support. Meanwhile, it considers the future mining of the integrity between the coordinated exploiting of coal and associated resources and disaster prevention and controlling. It mainly involves coordinated exploiting of coal and unconventional natural gases, coal and oil, and coal and water. Exploiting coal and associated resources is beneficial for realising safe and highly efficient recovering of multiple resources, and improving the resource utilisation rate. It can provide a brand-new thinking for the exploiting mode of resources in the future.

V Development direction of mining technologies in China

(1) Geological guarantee of coal resource exploitation. Delicate exploration in coal basins and research on multiple source geological disaster detection technologies can be conducted. The accuracy of geological exploration in coal basins in areas such as high Gobi Desert, loess tableland and gobs should be improved. In underground coal mines, the all-dimensional and visualisation predictable ability for multiple source geological disasters should be enhanced. The predictable forecast in advance and prevention ability for coal mine disasters should be strengthened, providing technical support for realising safe and efficient exploitation.

(2) Construction of large-scale mine sites at special strata. Emphasis is focused on key technology and equipment development such as deep vertical shaft with thick alluvium in the east, inclined shaft freezing in the west, drilling and mechanised construction in weak rock, comprehensive rapid drilling with full cross-section in stone drift, fulfilling the rapid construction requirement of large diameter mine sites.

(3) Efficient exploitation of coal resources. Emphasis should be focused on the new materials of key components of fully mechanised equipment, new techniques, systematic match, warm boost of large-scale equipment, rapid transport, controlling and reinforcement of deep surrounding rock masses in coal mines, and research on technical equipment in mechanised exploitation of coal seams with extremely thin thickness. Research on water-preserved mining and ecological restoration should be conducted. The complete set of efficient backfilling equipment should be developed, realising exploitation with low damaging. Large-scale open-pit coal mine equipment and new high-power anti-explosion transport equipment should be developed.

(4) Coal mine informatisation and management modernisation. Emphasis should be focused on

research on the key technology of coal mine internet of things. Key technologies and equipment in comprehensively mine monitoring with high reliability should be developed, promoting intelligent controlling and modernisation of decision-making system. The key technology in electronic commerce which can promote the management informatisation and modernisation of coal mine companies. Additionally, key technologies such as training of coal mine employees should be conducted.

(5) Exploitation of coal and coal seam gases. The accumulation law of coal seam gases should be studied intensively. Technical equipment and techniques including efficient exploiting coal seam gas, integrated exploitation of coal seam gas and coal resources, and utilisation of coal seam gases with low concentration should be studied. Technologies including horizontal shaft with multiple branches, stabilisation of shaft sides, continuous layered fracturing technology, concentrated transport technique of coal seam gases with low pressure and monitoring technology should be performed. The coal seam gas exploitation and coal mining engineering should be combined tightly. The harmonious development mode between coal seam gas under different condition and coal exploitation should be constructed, providing advanced technology and equipment for industrialised exploitation of coal seam gases.

(6) Prediction and prevention of coal mine disasters. Research should be conducted in the following aspects, including the coupling disaster-induced mechanism for the dynamic hazards of coal and rocks with multiple factors, and detection analysis of forewarning information and monitoring warning technology. Key technical equipment such as eruption of coal and gases, comprehensive prevention of rock burst and detection of hidden fire sources should be developed. The system of coal and rock dynamic disaster prevention technology should be completed, promoting the comprehensive prevention level for coal mine dynamic disasters.

(7) Safely avoiding the danger in coal mines and emergency rescue. Underground movable large-scale refuge equipment with perfect function and reliable performance should be constructed. The series of facilities of fixed refuge chambers and personal protection equipment should be developed. The environmental parameters in the disaster area and the signals of voice and video should be transported stably. The standard system which can adapt to the coal mine disaster risk management in China and hierarchical control should be established.

(8) Prevention of coal mine occupational hazards. Dust prevention key technologies mainly including respirable dusts and the research on key technologies such as dust forewarning, dust concentration sensor and dynamic monitoring. With the mining depth increasing, the mine geothermal temperature continuously increases. Furthermore, heat releasing of mechanical equipment has already become one of important heat sources. The number of heat damage mine sites increases obviously. Considering the issues including low integration between current cooling technology and equipment, low efficiency in cooling and high operating cost, it is urgently required that research on technologies including prediction and prevention, and equipment on deep mine sites should be conducted.

(9) Coal processing and conversion. Research on underground coal preparation techniques and its set of technologies and equipment should be conducted.

(10) Energy conservation and emission reduction, and full utilisation of resources. Emphasis should be focused on developing efficient separation and concentration technology for coal seam gases, promoting the efficiency of coal burning and developing technical equipment for sulphur fixing and removal. Research on mine water treatment in the underground, large-scale cyclic utilisation, treatment of waste coal chemical water and technical equipment for efficient utilisation of mine thermal energy should be conducted.

VI Contents, objectives and learning methods of new mining technologies

This course aims at systematically introducing the new mining technologies, new progress on mine pressure and its controlling and new scientific technologies to solve mine roof disasters and dynamic pressure issues. In the teaching process, this course emphasises letting students have comprehensive understand and grasping of new mining technologies. This not only expands new knowledge for students but also provides basement for students to reasonably apply new mining technologies to analyse and solve complicated engineering and quality issues in mining engineering.

Through learning this course, the following objectives should be achieved.

(1) Understanding the mining industry and scientific development outline;

(2) Grasping new mining technologies in terms of top coal caving, large mining height, thin coal seam and extremely closed coal seams;

(3) Grasping domestic and overseas progress situation in mining and safety technology. Also, grasping new mining technologies in drilling mining and open-pit mining;

(4) Grasping forewarning of mine roof disasters, mine pressure and its controlling in large-scale fully mechanised face and mechanised caving face with large dip angle;

(5) Grasping research progress on new rock bolt and cable bolt reinforcement, controlling surrounding rock mass in the gob-side entry and pressure relief controlling of coal entries under high stress and weak rock condition;

(6) Grasping surrounding rock mass controlling technology in deep mining, management principles and practices of floor bump and progress together with prevention technologies in rock burst research;

(7) Being familiar with backfilling technology, mining technology with the replacement of waste gangue and intelligent mining technology;

(8) Understanding and prospecting the mining problems and key technology in exploiting minerals in ocean, polar region and highland.

The Modern Technology of Mining Engineering is a technical course that has rigorous theory and high systematic level. Before learning this course, students should study several fundamental courses such as Coal Mining and Mine Pressure and Strata Control. After grasping sort of foundation, learning this course then can digest and fully realise the contents, achieving the required learning effect. Besides, among the learning process, integrating theory with practices is also an approach to promote the learning efficiency. Through learning this course, students can grasp new technologies in mine exploitation and grasping the ability of analysing and solving mining engineering problems.

1 New technologies of mine exploitation in China

1.1 Top coal caving mining technology in extremely large-scale mines[3]

The top coal caving mining has several unparalle lsuperiorities such as the extremely large output with a single working face, the extremely high efficiency and the prominent benefit. Therefore, it has already become the main mining method for thick and extremely thick coal seams in China. With the construction of the large-scale modern mine sites and scientific technologies in China, mines develop towards automation, full mechanisation and large scale. The production technique that regards the large-scale fully mechanised working face those annual production is more than ten-million tons as the core, becomes the new development direction. Therefore, the top coal caving mining technology of the extremely large-scale mines is playing the important leading effect on the production and development of the coal industry in China.

1.1.1 Geological production conditions of the Tashan Coal Mine

The Tashan Coal Mine uses the mixed development method of surface shaft and vertical shaft. For the main transport system, the continuous transporting method with belt conveyor is used. The main surface shaft connects directly with the belt conveyor in the haulage roadway. There is no buffered coal bunker. The thickness of the main exploiting coal seam ranges from 11.1m to 31.7m (the average thickness is 19.4m). The variation amplitude is relatively large. There are 6 to 11 layers of gangues. The maximum thickness is up to 0.6m. The coal seam is invaded by igneous rocks. The coal seam and the roof are destroyed to different extents. The exploiting is quite difficult. In the local area of the coal seam, there is false roof. The rocks are carbon mudstones. The immediate roof is siltstones and carbon mudstones. Besides the coal seam 2#, the main roof is not apparent. The floor is mainly composed of carbon mudstones and siltstones. The length of the working face is 250m. The advancing length ranges from 3000m to 5000m. The Tashan Coal Mine is an extremely large-scale mine site whose annual coal output is 15 million tons. It has already reached the first-class level domestically in terms of the single output of the working face, per-capita efficiency, recovery rate of coal and profit-tax rate of the cost. It belongs to the modern high-production and high-efficient mine site.

1.1.2 Coal mining techniques

The mining method isa single working face along the strike direction with large mining height together with fully mechanised caving with the low caving position. The cutting height is 3.5m and the caving height is 10.03. The ratio between the cutting height and caving height is 1 : 2.86. At the initial mining stage of the working face, the top coal does not collapse or the collapsing of the top coal is not enough. To safely and efficiently recover the top coal, a cycle of two cutting together with one caving is performed. The advancing length with a cycle is 0.8m and the coal caving interval is 1.6m. After the first roof weighting occurs at the working face, a standard cycle of one cutting together with one caving is performed. Under this cycling condition, the coal caving interval is 0.8m. As for the roof, the natural roof collapsing method is used to manage the roof above the gob.

The production technique is: oblique cutting of the coal shearer → coal cutting → pushing the front scraper conveyor → coal caving → pulling the rear scraper conveyor.

The coal shearer uses the coal cutting method along two directions, specifically from the head direction to the tail direction and then from the tail direction to end head direction. Along the traction direction, the front drum cuts the top coal while the rear drum cuts the bottom coal.

1.1.3 Equipment configuration

The selection of the fully mechanised caving mining equipment is as following:

(1) Coal shearer. The fully mechanised caving mining is used and in the coal seam, the quantity of mixed gangue is relatively large and the strength of it is relatively high. Therefore, the SL500AC coal shearer developed by the Germany Eickhoff Company is used, as shown in Figure 1-1. The maximum production ability of this coal shearer is 2700t/h.

Figure 1-1 Coal shearer

(2) Flexible scraper conveyor. The transport ability of the scraper conveyor should be able to meet the requirement of safety and efficiency at the working face. The PF6/1142 front scraper conveyor and PF6/1342 scraper conveyor are used, as shown in Figure 1-2. The transport ability of them reaches 2500t/h and 3000t/h.

(3) Transfer machine and crusher. At the end of the haulage roadway of the working face, the PF6/1542 transfer machine is arranged. The transfer machine delivers the coal blocks coming from the front scraper conveyor and the rear scraper conveyor to the belt conveyor. The transferring

Figure 1-2　Flexible scraper conveyor

capacity is 3500t/h. In the haulage roadway of the working face, a crusher is also arranged. Specifically, the crusher is installed on the top of the shute of the transfer machine. The Sk1118 crusher is adopted and its crushing ability is 4250t/h.

(4) Hydraulic support. Currently, the top coal caving hydraulic support mainly has two types, in which one is the reverse four-bar linkage top coal caving support while the other is conventional four-bar linkage top coal caving support. The advantage of the reverse four-bar linkage is that the latter space is relatively large. Therefore, it is more convenient for caving the top coal and maintaining the rear scraper conveyor. On the other hand, the shortcoming is that under the condition of high resistance, the stability and strength of the four-bar linkage structure is restricted. Compared with that, the conventional four-bar linkage structure can overcome the shortcoming occurring in the reverse four-bar linkage hydraulic support, which is widely used in China. Nevertheless, the shortcoming of the conventional four-bar linkage hydraulic support is that the latter maintenance space is relatively small. In China, the traditional four-bar linkage hydraulic support reaches a record of 6 million tons of output in one year in a single working face. Finally, based on the geological characteristics of the coal basin, the traditional four-bar linkage top coal caving hydraulic support with low caving position is used, as shown in Figure 1-3. The main technical parameters of kinds of hydraulic supports are tabulated in Table 1-1.

Figure 1-3　Hydraulic support

1.1 Top coal caving mining technology in extremely large-scale mines

Table 1-1 Technical parameters of hydraulic supports

Name	Traditional hydraulic support	Transition hydraulic support	Hydraulic support at the working face end
Type	ZF/13000/25/38	ZF/13000/26.5/38H	ZTZ20000/25/38
Initial support load/kN	10096	10096	15464
Working resistance/kN	13000	13000	20000
Height/mm	2500~3800	2500~3800	2500~3500
Length×Width/mm×mm	5395×1750	6055×1750	11755×3340

(5) Extension-type belt conveyor in the haulage roadway. The belt conveyor in the haulage roadway is selected based on 1.6 times of the coal cutting ability at the working face. Then, the DS1400/3300 belt conveyor is selected. Its transport ability is 3000t/h and the belt speed is 4.5m/s. The belt width is 1.4m.

Besides, the EHP-3K200/53 hydraulic pump is selected. Its production ability is 309L/h. The EHP-3K125/80 and KMPB320/23.5 atomising pumps are used. And their production ability is 516L/h and 320L/h respectively. Furthermore, transferring vehicles which are used to move hydraulic supports are arranged, as shown in Figure 1-4.

Figure 1-4 Transferring vehicle for hydraulic supports

1.1.4 Roof management

At the working face, 126 ZF/13000/25/38 traditional four-bar linkage mechanised caving supports are used and 7 ZF/13000/26.5/38H transition hydraulic supports are used. Additionally, ZTZ20000/25/38 hydraulic supports that are located at the end of the working face are used to support the roof. The traditional natural roof collapse method is used to manage the roof. The center-to-center distance of the hydraulic support is 1750mm. The maximum roof controlling distance is 6564mm while the minimum roof controlling distance is 5764mm. The cutting plane distance is controlled within 369mm.

1.1.5 Ground pressure observation

(1) Roof separation monitoring in the roadway: The roof separation warning instrument developed

by the Shandong Uroica Company is used. The data measured by the roof separation warning instrument is collected manually by the staff in the Ground Pressure Unit of the Production Technology Department.

(2) Monitoring of the advanced abutment pressure: At two ends of theworking face, for the advanced support, the single hydraulic prop is used to support the roof. Due to the fact that frequent installation and uninstallation of the single hydraulic prop occur with the working face moving, it is not applicable for online monitoring. Therefore, the working resistance recorder for single hydraulic prop manufactured by the Shandong Uroica Company is used. The data is recorded manually with the portable data collector.

(3) Monitoring of the hydraulic support working resistance: The ZVDC-1 computer monitoring system for fully mechanised supports manufactured by the Shandong Uroica Company is used. The pressure instrument is installed on the underground hydraulic supports, transferring the recorded pressure data to the ground computer system with the cable line to conduct data analysis. This realises continuous online monitoring.

1.1.6 The high production and high efficiency mode of Tashan Coal Mine

The coal mines of Tong Coal Group mainly exploit the coal seam belonging to the Jurassic System. As a consequent, there is always no experience in exploiting coal seams belong to the Permo-Carboniferous system. Therefore, a large number of technical problems occurring in recovering the extremely thick coal seams belong to the Permo-Carboniferous system with high safety, high efficiency and high resource recovery ratio. As the initially constructed mine of the Tong Coal Group when exploiting the coal seams belonging to the Permo-Carboniferous, the Tashan Coal Mine mainly exploits the coal seam with a thickness ranging from 11.1m to 31.7m (the average thickness is 19.4m). The variation scale of the coal seam thickness is relatively large. 6~11 layers of mixed gangues are included in the coal seam and the maximum thickness is up to 0.6m. Many parts of the coal seam are invaded by the igneous rocks. The coal seam and the roof are destructed by the invasion of the igneous rocks to some extents. Consequently, it is pretty difficult to conduct mining activities. The constructed Tashan Coal Mine adopts the top coal caving mining method which is advanced domestically and overseas. The resource recovery ratio is more than 85%. The coal mining equipment is the most advanced high-power coal shearer in the world, guaranteeing the reliability of the equipment operation. Furthermore, modern safety measurement and monitoring system is equipped, which conducts dynamic monitoring on the staff, environment and equipment located not only in the underground but also on the ground surface, such as the coal preparation plant. With this method, the remote centralised control is realised. During the construction process of extremely large-scale mines, the Tashan Coal Mine preliminarily explores the safe and efficient mining technologies, which include:

(1) Mine exploitation mode. The Tashan Coal Mine adopts a mixed development mode, namely the horizontal adit together with the vertical shaft. Parallel excavation is conducted in the main horizontal adit and the auxiliary horizontal adit until it reaches the strata 3# to 5#. In the main

transport system, the continuous transport mode of belt conveyor is used. Between the main horizontal adit and the transport entry, the belt conveyor is used to connect them directly. Therefore, there is no buffering coal bunker. The number of determined coal mine employees is 584 and the whole staff efficiency is 152.9t/person. Coal mining equipment and mining techniques which are advanced domestically and overseas are used in this mine site. The highpower coal shearer and corollary equipment for the working face are equipped. The auxiliary transport system of trackless tyred vehicle is imported. The continuous coal shearer is used to excavate roadways. High power belt conveyor is used to transport the coal blocks intensively. The mining technique in this mine site uses the advanced technique of cutting the total height (5m) in a single time. The length of the working face is 250m. As for the roadways, the length is ranged from 3000m to 5000m.

(2) Technical equipment. When the Tashan Coal Mine conducts collection on the equipment, they fully consider the capacity in each link when themine production releases. Equipment in the working face such as the highpower coal shearer which is advanced domestically and overseas, and other auxiliary corollary equipment are used, which efficiently guarantees the increasing of mine output. After this mine goes into operation, the yearly output of a single working face for a single team reaches ten million tons. The output for a single month is more than 1310 thousand tons and the output for a single day is 58 thousand tons. Engineering practices demonstrate that the reliability of the equipment ability is the key in guaranteeing the continuous stability and high efficiency. In the same time, the Tashan Coal Mine pays attention to the importation of equipment. They collaborate with research institutions, universities and manufacturers jointly to develop fully mechanised caving mining equipment which realise full localisation for large mining height and the level of ten thousand tons.

(3) Support technologies. The wide expansion and application of different kinds of entry rock bolts and steel mesh reinforcement technologies are used. For example, during the excavation and advancing in the rock entry, utilisation of rock bolts is realised. For half-coal roadways or coal roadways, the rock bolt reinforcement ratio is improved. Furthermore, research and promoted application of steel mesh support technology in the areas such as fault, folds and fractured zones, provides security for conducting fast tunnelling in safe and high efficient mines.

1.2 Large mining height technology in thick coal seams[4]

Coal is the principle energy in China. Meanwhile, thick coal seam is the principle exploiting coal seam in 13 coal bases with the scale of 100 million tons. Exploiting coal resources with safety, high efficiency and high recovery ratio is the important industrial policy of the state. Meantime, it is the objective that coal companies pursue. The apparent advantages of fully mechanised mining with large mining height include high production, high efficiency, high security and high recovery ratio. Therefore, it becomes the important development direction of the thick coal seam exploiting technology. In this section, the Shendong Shangwan Coal Mine is regarded as an example, to illustrate the large mining height coal mining technology in thick coal seams.

1.2.1 Geological production conditions of the Shangwan Coal Mine

The Shangwan Coal Mine uses the combined development method with inclined shaft-surface shaft-vertical shaft, as shown in Figure 1-5. In coal seam 12#, the Panel 4# is located in the middle of the mine site. The thickness of the coal seam in this panel ranges from 7.93m to 9.68m. The dip angle of the coal seam ranges from 1 to 3 degrees. And this coal seam belongs to stable coal seam. The fully mechanised working face 12401 is the initial exploiting working face of the Panel 4# in coal seam 2#. The thickness of the overlying rock strata ranges from 124m to 244m. The thickness of the loose bed ranges from 0 to 27m. The immediate roof of the coal seam is the sandy mudstone with the thickness ranging from 1.2m to 2.5m. The main roof is the siltstones with the thickness ranging from 5.68m to 20.34m. The floor is the mudstones with the thickness ranging from 0.96m to 1.29m. The length of this fully mechanised working face is 299.2m and the advancing length is 5254.8m. The designed mining height is 8.8m. The large mining height method is used. This mining method has the advantages of high resource recovery rate, high production efficiency and economic benefits. Furthermore, the Shangwan Coal Mine has abundant production experiences in exploiting coal seam with a thickness of 7m. Based on the reserves condition of the coal seam and the production technology, the large mining height with a thickness of 8.8m is the optimal scheme.

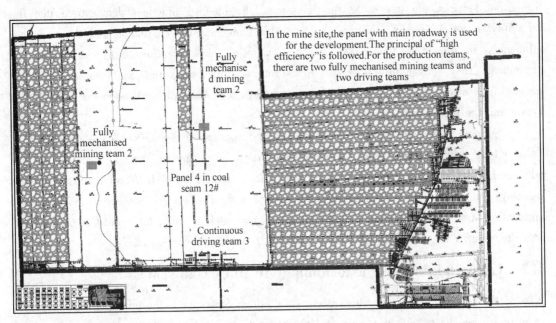

Figure 1-5 The mining and tunnelling layout of the Shangwan Coal Mine

1.2.2 Optimisation of the production system in Panel 4# in coal seam 12#

1.2.2.1 The auxiliary surface shaft

For the covered channel section of this surface shaft, the designed excavated cross-section is 7.2m ×

7.27m (width and height). It belongs to the rock tunnel with extremely large cross-section that is tunnelled in the highly weathered bed rocks. For the highly weathered bed rocks, the roof is fractured and the reinforcement is difficult. The mine uses multiple support and reinforcement methods to improve the support strength, such as the sheds without legs, rock bolt and meshes, and timbers, as shown in Figure 1-6. Through multiple technical methods to optimise the construction techniques, such as reducing the cyclic advancing length and tunnelling along different layers, the safe and high-efficient construction of the auxiliary surface shaft 2# is guaranteed. In the tunnelling process of the auxiliary transport surface shaft 2#, the worldwide problem of tunnelling the rock tunnels with extremely large cross-section in the highly weathered bed rocks is creatively solved. This provides the road for the entering of the fully mechanised mining equipment in the panel 4# in coal seam 12#. Furthermore, for the latter retreating of the equipment in the fully mechanised working face 12305, the transporting distance is reduced by 7km. This largely improves the moving efficiency of the panel 3# in coal seam 12#.

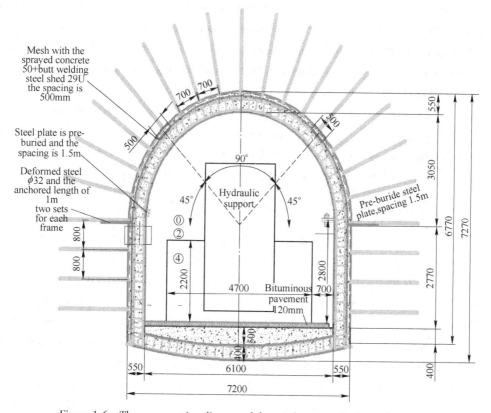

Figure 1-6 The cross-section diagram of the reinforcement in the surface shaft

1.2.2.2 Hydraulic support especially used roadway

For the fully mechanised working face of the Shangwan Coal Mine, the layout form of the roadway is the traditional panel with the single wing, as shown in Figure 1-7. In the fully mechanised working face in the panel 4# in coal seam 12#, the mining height is large. All retreat mining

Figure 1-7 Working sketch of the hydraulic support especially used roadway

equipment belong to the large-scale equipment. Additionally, when the horizontal auxiliary haulage roadway with large cross-section is regarded as the ventilation roadway of the next fully mechanised working face, the reinforcement strengthening of the roadway is massive. When it is influenced by the twice mining, the ground pressure appearance is severe, which is not beneficial for maintaining the roadway. Based on the existing problems in the layout of the traditional panel, the Shangwan Coal Mine make a breakthrough on the traditional roadway layout and optimised the layout of the production system. In the middle of the panel 4# in coal seam 12#, a support especially used roadway is excavated. It can provide installation and retreating service for 12 fully mechanised working faces in panel 4#. Furthermore, it can be used as the horizontal ventilation roadway for the last fully mechanised working face. Therefore, a roadway can be used for multiple times. In fact, it is an innovation of the working face moving and transporting system.

1.2.2.3 Open-off cut of the fully mechanised working face

To fulfil the requirement on the installation of the equipment in the working face, the maximum cross-section of the open-off cut 12401 is 14.4m×6.3m. The maximum span between the external corner of the installation joint roadway and the coal pillar of the working face is 15.6m. It belongs to the coal roadway with extremely large cross-section area that is tunnelled in extremely thick coal seam. The supporting and reinforcing requirement is high. Due to the fact that the designed height of the open-off cut is up to 6.3m. The current tunnelling equipment cannot fulfil the requirement of tunnelling the open-off cut with large cross-section area at one time. Therefore, it is determined that at the first time, the tunnelling height is 4.7m. Then, in the second time, the bottom section with a height of 1.6m is tunnelled. Then, the designed height of 6.3m is reached. The formation of the open-off cut is composed of two parts, namely the tunnelling and the excavating the bottom section. The open-off cut with the width of 11.4m is composed of two tunnelling cross-sections, namely 6m and 5.4m. The tunnelling cross-section with a width of 6m is always in front of the tunnelling cross-section with a width of 5.4m for 30m. Tunnelling and reinforcing are conducted alternately between those two cross-sections. Between the circulation, rock bolt reinforcement and cable bolt reinforcement should be conducted immediately. Only after all rock bolts and cable bolts are reinforced properly in the tunnelling face, can initiate the tunnelling for the next loop, as shown in Figure 1-8.

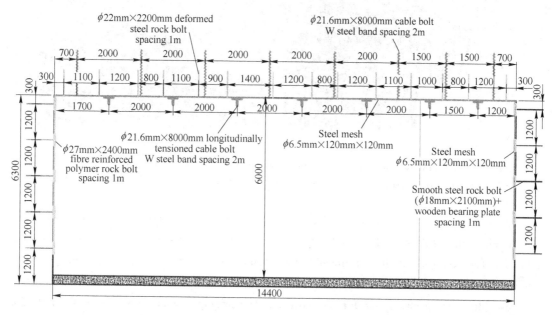

Figure 1-8　The reinforced cross-section of the corner of the hydraulic support at the end of the open-off cut 12401

1.2.3　Typical example of the set of fully mechanised working face equipment for large mining height (8.8m)

The Shendong Coal Group aims at the international advanced level, tightly cooperating with domestic and overseas coal mine equipment development manufacturers. They actively boost the localisation of the advanced coal mine equipment. Multiple new equipment and technologies are applied in the first fully mechanised working face with extremely large mining height of 8.8m in China. The set of fully mechanised mining equipment with extremely large mining height is shown in Figure 1-9.

(a)　　　　　　　　　　　　　　　　　(b)

(c)   (d)

Figure 1-9　The set of fully mechanised coal mining equipment for extremely large mining height of 8.8m
(a) Coal shearer; (b) Scraper converyor; (c) Belt conveyor; (d) Delivery vehicle for transporting hydraulic supports

(1) In the fully mechanised working face with extremely large mining height of 8.8m, the width of the belt in the haulage roadway is 1.8m. The newly developed strainless steel carrier roller is adopted. The single-point driving transporting distance is 5340m. The company independently studied and developed the new belt conveyor, which increases the transportation efficiency and decreases the power consumption for one-ton coal. The principal parameters include the following aspects that the power is 3×1600kW. The width of the belt is 1800mm. The length of the belt conveyor is 6000m. The moving velocity of the belt is 4.5m/s. The transportation ability is 4500t/h. The type of the speed reducer is WH3SH300 with the water cooling used.

(2) The Shendong Company independently studied and developed the self-moving train. With the process of advancing, disassembling the railway lines and assembling the railway lines are not necessary any more. This is beneficial for improving the level of automation and decreasing the labour strength for the employees. The external dimension of this self-moving train is 300000mm× 1400mm × 5500mm. It is applicable for the inclination angle ranging from 0 to 7 degrees. As for the height of the entry, it is applicable for the entry whose height ranges from 1850mm to 5500mm. The stepping distance is 2m.

(3) Then Shendong Company and the Shanghai Branch Company of the Tiandi Science and Technology jointly studied and developed the coal shearer with a height of 8.8m, which has the largest mining height and largest efficiency in China. The principal parameters include that the overall power is 2925kW. The height of the coal shearer is 4.2m. The mining height ranges from 4.3m to 8.6m. The weight of the coal shearer is 220t. The production capacity of the coal shearer is 6000t/h.

(4) The Shendong Company, the Zhengzhou Coal Mining Machinery (Group) Cooperation Limited and the Jiangsu Tianming Company jointly developed the hydraulic supports with the height of 8.8m and three equipment in the working face. Among them, the principal parameters of the hydraulic supports include that the support area ranges from 4m to 8.8m. The working resistance of the hydraulic supports is 26000kN. The supporting strength of the hydraulic supports ranges from 1.7MPa to 1.8MPa. The weight of the hydraulic supports is 100 tons. The center-to-

center distance of the hydraulic supports is 2400mm. The length of the plate that is used to support the coal wall is 4130mm.

(5) To fulfil the requirement that the hydraulic supports with a height of 8.8m and the coal shearer will be moved to other working faces, the Shendong Company cooperated with the Aerospace Heavy Industry Corporation Limited, jointly developing the delivery vehicle which can move a hydraulic support with the weight of 100 tons and the electromobile. Among them, the principle parameters for the delivery vehicle which is used to move hydraulic supports include that the length, width and height of the delivery vehicle are 8800mm, 4500mm and 2730mm respectively. The self-weight of the delivery vehicle is 50 tons. And bearing capacity of the delivery vehicle is 100 tons.

1.2.4 Innovative outcomes

In the construction process of the fully mechanised working face with extremely large mining height of 8.8m, the instrument for guiding the gangues and lowering dusts, instrument for adjusting the self-moving train and the instrument for injecting oil in the coal shearer automatically are developed chronologically. This improves the operating environment of the operators and the production efficiency, as shown in Figure 1-10. Besides, this optimises action setting of the flank oil cylinder in the hydraulic support. Therefore, the production technical problems such as the falling of the supports in the hydraulic working face and the trouble between the pushing and dragging rod of the hydraulic supports and the main leg.

Figure 1-10 The innovative equipment in the fully mechanised working face
with extremely large mining height of 8.8m

Through the ground pressure observation, the Figure 1-11 can be drawn. It can be acquired that in the range of 300m in which the fully mechanised working face 12407 advances, the ground pressure appearance is relatively severe. In the periodic weighting of the working face, the opening rate of the safety valves is approximately 25% and the maximum weighting strength is 520 bar. Through statistics, in the periodic weighting period, the decreasing value of the hydraulic support vertical column is up to 30mm for each coal cutting. The step of the periodic weighting ranges from 9m to 15m. After the working face advances for 300m, the ground pressure appearance is relatively relieved and the maximum weighting strength is 510 bar. The step of the periodic weighting is 16m to 19m. In the weighting period, the opening rate of the vertical column safety valve (475 bar) is approximately 20%.

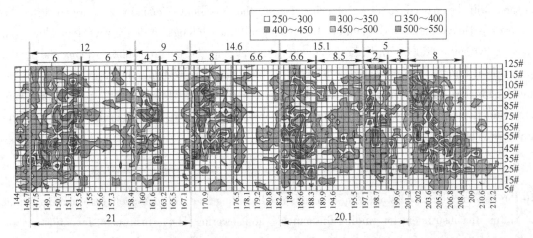

Figure 1-11   Ground pressure observation diagram

For the fully mechanised working face with extremely large mining height of 8.8m, the maximum production efficiency is 58.4 thousand tons per day. And the maximum monthly production ability is 1460 thousand tons. For a single working face, the annual coal production is 13 million tons. In the construction process of the fully mechanised working face, new technologies and new equipment, such as the microearthquake and the ground pressure monitoring technology, dynamic monitoring of the mining height and the stress state of the hydraulic supports in the fully mechanised working face are applied. This largely improves the safety and the intelligent level of the fully mechanised working face.

1.3   The fully mechanised coal exploiting technology in thin coal seams

Thin coal seam exploiting has several special problems, for example, the thickness of the exploited coal seam is thin and the working condition is poor. Furthermore, moving the equipment is difficult. The variation of the coal seam thickness is high. The ratio between the input and the output is high. Therefore, the economic benefits of the coal seams are less than thick and medium thick coal seams. From a certain extent, exploiting the thin coal seam has already seriously influenced the coordinated development of the whole coal mining industry. How to realise the high efficient mining of the thin coal seams under the condition of localisation of the fully mechanised coal mining equipment, is the problem that needs to be solved urgently. This section takes the Laoshidan Coal Mine as an example, illustrating the fully mechanised mining technology in the thin coal seam.

1.3.1   Geological production condition of the Laoshidan Coal Mine

In the Laoshidan Coal Mine, the average coal seam thickness of the working face 1208 is 1.2m, belonging to thin coal seam. The average dip angle of the coal seam is 8 degrees and it belongs to stable coal seam. Above the coal seam, there is a layer of mudstone with the thickness ranging from 0 to 0.10m. The immediate roof is sandy shales with a thickness of 1.8m. The main roof is fine sandstones with a thickness of 10m. The immediate floor is fine sandstone with a thickness of 1.88m. The main floor is sandy shale with a thickness of 7.11m. The length of the working face is 150m and the advancing length is 800m. The fully mechanised coal mining method to cut the coal seam with a single cut is used. The thin coal seam mining is slow, making that the medium thick coal seam in the bottom section cannot be exploited in time. This seriously influences the connection problem in exploiting of the several coal seams in the Laoshidan Coal Mine.

1.3.2 Technical design and equipment selection of the fully mechanised mining in the thin coal seam

1.3.2.1 Overall design of the fully mechanised technique in the thin coal seam

According to the reserve status of the coal seam 12# in the Laoshidan Coal Mine of the Haibowan Mining Company, considering the matched mining and the stubble repeating of the whole mine, the fully mechanised longwall working face along the strike direction is used in this coal seam. To improve the output of the thin coal seam working face in the Laoshidan Coal Mine, the production ability of this set of equipment is determined as 0.3~0.4 million tons/a. It is the best to arrive at 0.4 million tons/a. To fulfil this target, it is determined that the length of the working face is 150m and the average mining height is 1.2m. The advancing length in a cycle is 0.6m. The driving velocity of the coal shearer is more than 6m/min. The recovery rate of the coal resources reaches 98%. The recommended parameters for the working face are tabulated in Table 1-2.

Table 1-2 Determination of the main parameters of the working face

Name	Value
Length of the working face/m	150
Mean mining height/m	1.2
Cutting depth of the coal shearer/m	0.6
Coal recovery ratio/%	98

The coal shearer cuts the coal seam along the inclined direction. The cutting position is located in the middle of the working face where in the roof is favourable. The length of the coal cutting is 20m. For the moving of the hydraulic supports and the pushing the scraper conveyor, operations following the coal shearer are used. Its technical process is: coal cutting-moving of hydraulic supports-pushing the scraper conveyor. When the coal shearer cuts the coal with a small velocity, for the hydraulic support moving method, the successive moving of the single hydraulic support is used. When moving of hydraulic supports is behind the coal shearer for a large distance, the stagger moving of hydraulic supports with different groups is used.

1.3.2.2 Equipment selection

A Selection design of hydraulic supports

(1) Selection principles of hydraulic supports, as shown in Figure 1-12.

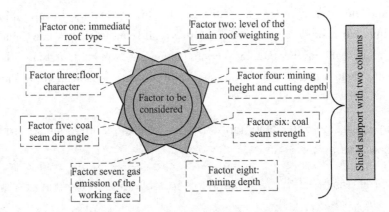

Figure 1-12 Parameters that need to be considered in the selection of hydraulic supports

(2) Determination of hydraulic support parameters.

Based on calculating and the selection principle, it is determined that the fully mechanised working face in this thin coal seam uses theshield supports ZY2600/07/16. For the end of the fully mechanised working face, the hydraulic supports ZYG3200/12/25 that are matched with the hydraulic supports ZY2600/07/16 are used. The main technical parameters of the hydraulic supports in this coal seam are tabulated in the following table. The main technical parameters of the hydraulic supports in the coal seam are shown in Table 1-3 and Table 1-4.

Table 1-3 Main technical parameters of the middle hydraulic supports

Model number	ZY2600/07/16
Type	Shield supports
Height/mm	700~1600
Width/mm	1470
Width between centres/mm	1500
Initial supporting force/kN	2182
Working resistance/kN	2600
Supporting strength/MPa	0.36~0.43
Pressure on the floor/MPa	<2
Pressure of the pump/MPa	31.5

Table 1-4 Main technical parameters of the hydraulic supports at the end

Model number	ZYG3200/12/25
Type	Shield supports
Height/mm	1200~2500
Width/mm	1470
Width between centres/mm	1500
Initial supporting force/kN	2618
Working resistance/kN	3200
Supporting strength/MPa	0.45~0.5
Pressure on the floor/MPa	<2
Pressure of the pump/MPa	31.5

B Selection design of the coal shearer

(1) Selection principle of the coal shearer.

Selection of the coal shearer should consider the mining geological condition of the coal seam, the practical production ability of the working face, the designed production ability of the working face, the matching ability of the mine system, and the supplying status of the equipment. When the geological condition in the coal seam exploiting is favourable, the matched ability of the mine system is large and the designed production ability of the working face is large, the coal shearer with large power can be selected based on the situation. When the geological condition of the coal seam is complicated and the ability of the mine system is small, the coal shearer group that requires smaller investment can be selected.

(2) Determination of the parameters for the coal shearer.

According to the reserve features of the thin coal seam and consideration of the thickness of the hydraulic support top beam, sinking of the roof, bumping of the floor, the minimum mining height, matching between the coal shearer and hydraulic supports, and the minimum safe space to move the machine, it is preliminarily determined that the coal shearer driven by the alternating current used in thin coal seam MG100/238-BWD that is developed by the Brand in Shanghai of the Tiandi Science and Technology Corporation is used. The parameters are shown in Table 1-5.

Table 1-5 Main technical parameters of the coal shearer

Model number	MG100/238-BWD
Cutting depth/mm	630
Diameter of the drum/mm	800
Range of the mining height/mm	950~1250
Hardness of the coal	$f \leqslant 4$
Voltage of the power supply/V	1140

The coal shearer MG100/238-BWD for thin coal seam is shown in the Figure 1-13. The space above the shovel is utilised. The method of longitudinal layout of the cutting motor is used. This successfully solves the contradiction between the height of the coal shearer, the space to move the coal and the installed power. For the cutting section, the longitudinal layout of the cutting motor is used. For the driving section, the horizontal layout of the driving motor is used. This finds an efficient approach to decrease the height of the coal shearer, for the design of the coal shearer for the thin coal seam. This layout form is beneficial for promoting the development of the coal shearer technology for the thin coal seam.

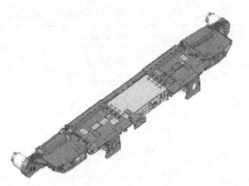

Figure 1-13　Coal shearer of MG100/238-BWD
(When the drum is not installed)

C　Selection design of the scraper conveyor

(1) Selection principle of the scraper conveyor.

First, the transport ability of the working face should be guaranteed. Furthermore, there should be a certain back-up for the scraper conveyor. Secondly, based on the in-situ situation, the driving mode of double motors and double heads should be preferred. Additionally, to match the requirement that the coal shearer needs to cut the coal along two directions repeatedly, the shovel plate should be installed along the one side of the scraper conveyor that is adjacent to the coal. The aim of it is to clean the floating coal in the route.

(2) Determination of the scraper conveyor parameters.

According to the calculation of the production ability of the working face and the existing designing experience, the scraper conveyor of SGZ-630/320 should be selected. The length of it is 150m and the transport capacity is 450t/h. The motor power is 2×160kW. The main technical parameters of the scraper conveyor for the thin coal seam are tabulated in Table 1-6.

Table 1-6　The main technical parameters of the scraper conveyor

Model number	SGZ630/320
Length after it is manufactured/m	150
Transport volume/$t \cdot h^{-1}$	450
Chain speed of the scraper/$m \cdot s^{-1}$	1.1
Dimension of the middle groove/mm×mm×mm	1500×590×195

D　Selection of the other main matched equipment

The matching of the main set of other equipment is shown in Table 1-7.

Table 1-7 The main coal cutting mechanical equipment used in the fully mechanised working face for the thin coal seam

No.	Name	Model number	Quantity	Unit	Power	Using place
1	Coal shearer	MG-238BW	1	Set	238kW	Working face
2	Scraper conveyor	SGZ-630/320	1	Set	2×160kW	Working face
3	Hydraulic support	ZY2600/07/16	95	Set	2182kW	Working face
		ZYG3200/12/25	5	Set	2618kW	Working face
4	Belt conveyor	DSP1040/800	2	Set	90kW	Haulage roadway
5	Tensioning winch		1	Set	8kW	Haulage roadway
6	Reversed loader	SZB-730/75	1	Set	75kW	Haulage dip entry
7	Signal fully protecting equipment	ZXZ8-2.5	2	Set	2.5kVA	Ventilation roadway
8	Emulsion pump	WRB125/31.5A	2	Set	2×75kW	Ventilation roadway
9	Dispatching winch	JD-11.4	4	Set	11.4kW	Ventilation roadway
10	Dispatching winch	JD-25	1	Set	25kW	Ventilation roadway

1.3.3 Reinforcement technology of the horizontal roadway in the thin coal seam

For the thin coal seam, the horizontal haulage roadway and the horizontal ventilation roadway are tunnelled along the floor of the coal seam. In the roadway, the rock bolt, the beam and the cable bolt are used for the reinforcement. The cross-section of the horizontal haulage roadway and the horizontal ventilation roadway are rectangular. The net width is 3.0m and the net height is 2.5m, as shown in Figure 1-14. Through simulation of the reinforcement scheme for the large cross-section roadway in the thin coal seam, it is acquired that the sinking value of the roof is 135mm and the deformation value of the roadway side is 118mm. The bump of the floor is 32mm. This indicates that this reinforcement scheme is reliable.

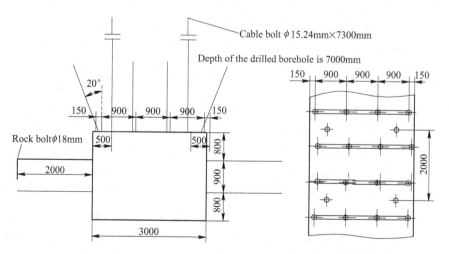

Figure 1-14 Layout diagram showing the reinforcement of the horizontal roadway 1208 in the Laoshidan Coal Mine

1.3.4 Ground pressure observation

The ground pressure observation shows that the practical initial supporting load of the whole working face is relatively small. The pump station, the hydraulic pressure pipe and detecting the malfunction of the hydraulic supports should be strengthened to improve the practical initial supporting load of the hydraulic supports. From the overall practically observed hydraulic pressure information of the hydraulic support vertical columns, when the main roof weighting comes, the opening frequency of the hydraulic support safety valves is quite small. The nominal working resistance of hydraulic supports can fully meet the requirement of the ground pressure controlling when the main roof weighting comes. The equipment selection in the fully mechanised working face for the thin coal seam is reasonable. The matching of "three equipment" is favourable. After the fully mechanised mining technology in the thin coal seam is used, the production of a single working face, the labour efficiency and the machine opening rate of the coal shearer are improved. This is beneficial for the concentrated production of the mine site. Furthermore, this has an important significance on extending the service life of the Laoshidan Coal Mine and realising the high-efficient mining.

1.4 Mining and safety technologies of the coal seams with extremely adjacent distance

In China, there are many mine sites in which the distance between the coal seams is extremely small. Due to the difference on the coal forming condition, the reserve condition of coal seams is different. Generally, it is more difficult to exploit the extremely adjacent coal seams. The difficult is mainly determined by the following parameters, namely the distance between up coal seam and the down coal seam, the thickness between the rock mass layers, the properties of rocks, the depth of the coal seam, the in-situ stress environment, the water filling condition and the gas content. How to adopt the corresponding technical measures under different mining environment, to improve the recovery ratio, extent the service-life of the mine site, and realise the safe and high efficient production is always the core problem that the coal mining industry concerns.

1.4.1 Geological production conditions of the Wuhushan Coal Mine

In the Wuhushan Coal Mine belonging to the Wuhai Energy of the China Energy, the coal seams 9# and 10# belong to extremely adjacent coal seams. The distance between the coal seams is usually ranged from 0.5m to 2.5m. The average thickness is 2.0m. In the coal seam, there are fractured mudstones. The joints and fractures are developed. The stability is extremely poor. Above the coal seam 9#, the immediate roof is composed of grey and black sandy clay mudstone. The average thickness is 3.5m. The main roof is composed of fine sandstones and the average thickness is 4.8m. The immediate floor is composed of sandy mudstones and the thickness is not uniform. The average thickness is 2.0m. For the working face 901, the length along the dip direction is 132m and the length along the strike direction is 440m. The structure of the coal seam 10# is simple and the fractures are relatively developed. The average thickness is 2.2m and the average dip angle of the coal seam is 7 degrees. For the floor, it is the fine sandstone with an average thickness of 5.4m. In the coal seam 9#, the working faces 901 and 903 finished mining in 2008. Therefore, the coal seam 10% which is below it, must be subjected to the complicated mining condition below the gob area of the coal seam 9#. Through classifying the stability of the roadway surrounding rock masses, the surrounding rock masses of the haulage roadway 1001 in the coal seam 10# belong to extremely instable surrounding rock masses (Level V). For this kind of roadways, the rock bolt reinforcement and cable bolt reinforcement cannot be only used. In fact, the metal support, as the main support components, should be used as the support form. The mining and tunnelling in coal seam 9# is shown in Figure 1-15.

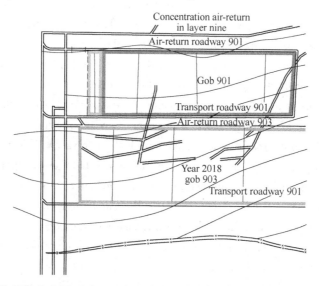

Figure 1-15 The horizontal diagram showing the mining and tunnelling in coal seam 9#

Based on the geological production condition of the coal seam 10# in the above-mentioned Wuhushan Coal Mine, the main difficulties in the exploiting of the extremely closed coal seam below the gob area of the Wuhushan Coal Mine.

(1) Between the coal seams 9# and 10#, there are mudstones. The strength is low. Being influenced by the exploiting of the coal seam 9#, joints and fractures are developed.

(2) The remaining coal pillars in the coal seam 9# seriously influences the determination of the proper position of the coal roadways in the below coal seam 10#.

(3) The roof thickness of the coal seam 10# is only 2.0m, which seriously restrains the support form. It is not beneficial for the roadway maintaining.

(4) Above the gob of the coal seam 9#, there are water-bearing strata. It is easy to lead to the hydrops in the gob area, which threats the exploiting of the coal seam 10#.

(5) When the coal seam 10# is being exploited, the roof of the fully mechanised working face is easy to be fractured. The coal rib spalling is serious and the ground pressure law is not clear.

(6) The coal seam 10# is easy to self-ignite. Furthermore, the gases in the gob areas of two coal seams erupt, leading to the high concentration of the gases in the gob area.

1.4.2 Determination and its controlling of the proper position of the mining roadways and the open-off cut

1.4.2.1 Study of the proper position of the mining roadways and the open-off cut

According to the practical geological production condition in the field, the numerical simulation software of $FLAC^{3D}$ is used, to determine the property position of the haulage roadway, the ventilation roadway and the open-off cut of the working face.

Aiming at the layout position of the haulage roadway of the working face 1001, 7 simulation schemes are designed (The distance between the central line of the haulage roadway and the edge of the coal pillar in the coal seam 9# is $-7.5m$, 0, 15m, 20m, 25m, 30m and 35m respectively), as shown in Figure 1-16. Analysis is conducted on its stress field, displacement field and the roadway failure range. It is acquired that when the haulage roadway is located below the coal pillar in coal seam 9#, due to the effect of the concentrated stress, it is easiest to generate the severe deformation. With the gradual increasing of the distance between the central line of the haulage roadway and the edge of the coal pillar in coal seam 9#, the influence of the concentrated stress of the coal pillar on the roadway deformation gradually decreases. When its distance is larger than 25m, there is no apparent influence. This simulation selected that the optimum scheme is 1-5, indicating that the distance between the central line of the haulage roadway and the edge of the coal pillar in the coal seam 9# is 25m. Based on the simulation, it can also acquire that the central line of the ventilation roadway of the working face and the edge of the coal face should be 21.5m. The scheme that the distance between the central line of the working face open-off cut and the edge of the gob area is 12m is optimum.

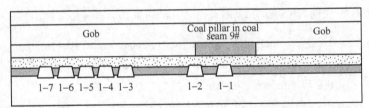

Figure 1-16 Simulation scheme to determine the proper position of the haulage roadway

The roadway arrangement form in exploiting of the lower coal seam in extremely adjacent coal seams mainly include four types, namely the internally staggered form, the externally staggered form, the horizontally staggered form and the overlapped form. Based on the above numerical simulation analysis, it is acquired that for the arrangement form of the mining roadways in the fully mechanised working face 1001, the horizontally staggered form should be used. Its superiority is reflected in that the stress concentration phenomenon resulted by the overlapped arrangement can be avoided. Furthermore, the phenomenon of coal resource loss that is resulted by the internally staggered arrangement form can be reduced. Additionally, it can address the problems that the arrangement of the adjacent working face is irregular which is resulted by the externally staggered form can be solved. The specific arrangement form is shown in the Figure 1-17.

1.4 Mining and safety technologies of the coal seams with extremely adjacent distance

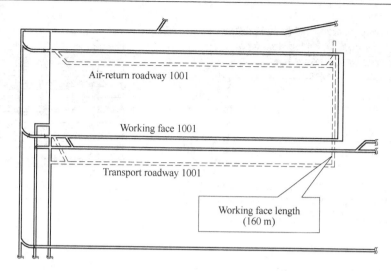

Figure 1-17 The horizontally staggered arrangement diagram of the mining roadways for the working face 1001

1.4.2.2 Study of the mining roadway supporting technologies

Above the roof of this roadway, it is the gob area. It is predicted that the roadway pressure will not be pretty large. Maintaining the integrity of the roof is the main objective of the controlling. Therefore, it is determined to use the passive support method. The metal support is regarded as the main support material.

The haulage roadway 1001 is along the roof of the coal seam 10#. Tunnelling is conducted along the floor. For the permanent support, the trapezoid U-shaped steel beam shed is used. As for the temporary support, the front canopy support is used. 4 steel tubes with the diameter of 2.5 inches and length of 2.5m are used as the front canopy beam. Steel tubes are inserted in the roof of the shed beam that has already been installed. Above the steel tubes, the wooden plates are used to insert tightly.

The ventilation roadway 1001 is along the roof of the coal seam 10#. The tunnelling is conducted along the floor. For the support form, the steel I-beam 11# + wooden beam + antiseptic mesh is used. As for the temporary support, below the steel I-beam, the flying rings are hung. In the flying rings, the steel tubes are regarded as the front canopy beam. On the front canopy beam, wooden plates are installed, to form the roof supporting mode of woods and wooden wedge. The specific supporting scheme is shown in Figure 1-18.

1.4.2.3 Reinforcement technology for the fractured roof in the extremely adjacent thin coal seams

For the roof of the fully mechanised working face 1001, the thinnest section has a thickness of 0.5m. Guaranteeing that when the open-off cut of the fully mechanised working face 1001 is being tunnelled, accidents such as the leaking and collapsing of the roof that may influence the production will not happen, has already become one of the important restraining parameters in

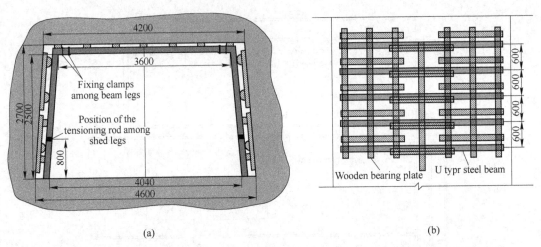

Figure 1-18 The supporting scheme of the haulage roadway in the working face 1001
(a) Trapezoid steel beam shed; (b) The projection drawing after the U-shaped steel beam shed is used

realising the favouring and safe installation of the equipment and exploiting in the working face 1001. The reinforcement requirement on the roof of the open-off cut of the fully mechanised working face includes conducting reinforcement on the roof before expanding the roadway. The rock masses in the roof and the floor can form an integrity. The reinforced rock masses have a long-term strength. The implementation is simple and it is easy to be grasped and extended.

A Selection of the grouting and reinforcement materials for the thin and fractured roof

The reinforcement technologies of the thin and fractured roof of the roadways with large cross-section in extremely adjacent coal seams mainly include the reinforcement technology with the concentration of the rocks and soils (such as the cement grout) and reinforcement technology with grouting chemical materials (such as the Marithan).

The main function of the reinforcing roof with chemical grouting is the function of the net framework of the grouting liquid induration, compressing to dense after backfilling and transferring the failure mechanism of the surrounding rock masses, improving the strength of the surrounding rock masses, forming the bearing structure and improving the reserve environment. Based on the performance of the above kinds of materials and the purpose, the Marithan is first used as the roof reinforcement materials. The classification of the grouting and reinforcing materials is shown in Figure 1-19.

B Design and practice of the thin and fractured roof reinforcement scheme for the open-off cut 1001

The shape of the open-off cut is the trapezoid cross-section. The tunnelling dimension is that the above width is 6.38m and the bottom width is 7.02m. The height is 2.8m. After the reinforcement is finished, the net dimension is that the above width is 5.8m and the bottom width is 6.44m. The height is 2.5m. The shape is tunnelled based on tunnelling with two times. During the first tunnelling process, after the tunnelling arrives at 70~120m, the roof thickness is less than 1.0m.

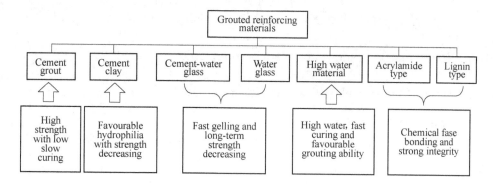

Figure 1-19 Classification of the grouting reinforcement materials

For the section 75~95m, the thickness is only around 0.4~0.6m. To guarantee that the roadway is stable in the service period of the open-off cut, it is determined that before the second expanding of the roadway, grouting and reinforcement is conducted on the surrounding rock masses. The specific scheme is shown as following:

The grouting drilled borehole has a diameter of 42 mm and the depth of the drilled borehole is $l = 3000$mm. The borehole drilling position is below the roof and the distance is $h_1 = 300$mm. The orientation of the drilled borehole is that it is perpendicular with the coal face and along the horizontal direction, the upward intersection angle is 12 degrees. The spacing between the drilled boreholes is 2500mm. The borehole sealing position is ranged from 1400mm to 1600mm. For the grouting pressure, the relatively lower value of 3~4MPa is selected. For each drilled borehole, the combined grouting volume is around 5 buckets.

In-situ practices show that the liquid of Marithan can fill the roof fractures properly, bonding the roof rock masses. It enhances the tensile strength of the roof and its integrity. In the second roadway expanding process, the accidents such as the roof collapsing and the rib spalling leaded by the damaging of the roof do not occur. And the construction period is not delayed. Consequently, pretty good effects are acquired. The grouting position and the parameter diagram of the Marithan are shown in Figure 1-20.

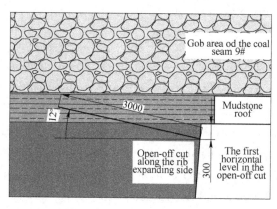

Figure 1-20 Grouting position of the Marithan and the parameters

1.4.3 Designing selection of the fully mechanised equipment

1.4.3.1 Determination of the parameters of the hydraulic supports

A Determination of the hydraulic support working resistance

Calculating method 1:
$$W = (6 \sim 8) \times h \times B \times \gamma \times \cos\phi = 2597.1 \text{kN/support}$$
Calculating method 2:
$$P_s = 72.3h_m + 4.5L_p + 78.9B_c - 10.24N - 62.1$$
$$Q = P_s \times b \times (B_c + B)/K_s = 3300.16 \text{kN/support}$$

Based on the calculating results of the above two methods, it is determined that the working resistance of hydraulic supports is 3800kN and the initial supporting force is 2660kN.

B Determination of the hydraulic support height

In the designed mining zone, the exploiting thickness of the coal seam is ranged from 1.5m to 2.8m. Therefore, for the maximum height of the hydraulic supports, $H_{max} = 2.8+0.2=3.0$m; For the minimum height of the hydraulic supports, $H_{min} = 1.5-0.2=1.3$m.

Based on the above parameters, it is determined that for the fully mechanised working face 1001, the ZY3800/14/30 type shield support is used.

1.4.3.2 Selection of the coal shearer

The selection of the coal shearer should consider the parameters including the geological condition of the coal seam exploiting, the practical production ability of the working face, the designed production ability of the working face, the setting ability of the mine system and the equipment supplying situation. Then, it is determined to use the coal shearer MG250/601-WD which is driven by the alternating current.

1.4.3.3 Selection of the scraper conveyor

First, the transport ability of the working face should be guaranteed. Furthermore, there should be a certain backup. Secondly, based on the in-situ situation, the driving mode of double motors and double machine heads is preferentially selected. Furthermore, to match the requirement that the coal shearer should cut the coal along two directions, the coal shovel plate is installed in the conveyor along the coal face side, to clean the floating coal in the channel. According to the calculation of the transport ability and referring to the designing experiences, the scraper conveyor SGZ-764/630 is selected. The length is 250m and the transport ability is 900t/h. The power of the electric motor is 2×315kW. The main coal cutting mechanical equipment in the fully mechanised working face is shown in Table 1-8.

1.4 Mining and safety technologies of the coal seams with extremely adjacent distance

Table 1-8 Principal coal mining equipment used in the fully mechanised working face

Number	Name	Type	Capacity	Quantity
1	Coal shearer	MG250/601-WD	600kW	1
2	Scraper conveyor	SGZ-764/630	315kW×2	1
3	Transfer machine	SZZ-764/160 self-moving	160kW	1
4	Crusher	PLM-1000	110kW	1
5	Self-moving machine	DY-1000	633kW	1
6	Belt conveyor	JSP-1080/1000	160kW	1
7	Emulsion pump	BRW-200/31.5	125kW×2	1 being used and 1 spare
8	Atomising pump	BRW315/16	110kW×2	1 being used and 1 spare
9	Movable electric substation	KBSGZY-1000/1140/660	1000kVA	2
		KBSGZY-630/6	630kVA	1
10	Drainage pump	BQK50-150	75kW	3
11	Air pump	QFB-70	30kW	2
12	Prop pulling hoist	JD-25kW	25kW	2
		JH-18.5kW	18.5kW	2
13	Hydraulic supports	ZY3800/14/30		111

1.4.4 Ground pressure observation and gas safety management in exploiting of the extremely adjacent coal seams

1.4.4.1 Observation of the in-situ ground pressure

The purpose of the ground pressure observation in the fully mechanised working face is collecting the information regarding the hydraulic support-surrounding rock masses, to determine the ground pressure state and the potential accident situation of the fully mechanised working face. This provides materials for the following work (such as the data analysing and processing, and proposing the measures to solve the accidents). The observed contents include the hydraulic information of the vertical columns of the hydraulic supports, the geometric state of the hydraulic supports and the roof collapsing and rib spalling of the working face. The ground pressure observation layout is shown in Figure 1-21.

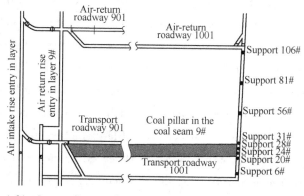

Figure 1-21 Layout diagram showing the ground pressure observation stations

The processing and analysing of the hydraulic information of the hydraulic supports are shown in Figure 1-22. It can be acquired that when the immediate roof of the working face initially collapses, the hydraulic pressure of the vertical columns in the working face show continuous higher for a short period. As for the rest normal retreat mining, the hydraulic pressure of the vertical columns in the whole working face distributes commonly lower. The peak of the hydraulic pressure occurs in the working face that is below the coal pillar. Furthermore, for the hydraulic supports at the coal pillar, distribution of the hydraulic pressure is wider compared with the top and the bottom. When the initial collapsing occurs, the distribution range of the relatively higher hydraulic pressure is larger. However, under the normal situation, only relatively scattered high hydraulic pressure occurs. For the immediate roof of the working face 1001, the average stepping of the initial collapsing is 15.2m. There is no apparent main roof weighting phenomenon.

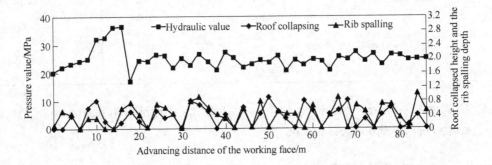

Figure 1-22 Hydraulic information of the coal pillar area in the working face 1001 and the changing characters of the roof collapsing and rib spalling

For the shield supports with two vertical columns in the fully mechanised working face 1001, the hydraulic pressure that is beyond the rated working resistance, accounts for 4.15%. The working resistance of hydraulic supports can fulfil the requirement of ground pressure controlling. For the geometric state of the hydraulic supports, in most zones, it can be controlled in the normal range. The working state is favourable. In a small range, the phenomenon such as the large top beam stairs, the large pitching of the top beam and the large distance to the coal face occurs. These mainly occur in the working face that is adjacent to the coal pillar in the upper coal seam. For most working faces, a small extent of roof collapsing and rib spalling occurs. For the observation stations 24#, 28# and 31# below the coal pillar, accidents such as roof collapsing and rib spalling are easy to occur. Therefore, in the working face, when coal cutting is near the coal pillar, the hydraulic supports should be pulled and advanced immediately to control the roof. This is to prevent the phenomenon of large roof collapsing which is resulted by the cutting through of the gob area of the coal seam 9# leaded by the roof collapsing. Overall, in the fully mechanised working face 1001, the shield support ZY3800/14/30 can meet the support requirement of the working face.

1.4.4.2 Gas safety management technology in exploiting the extremely adjacent coal seams

A Analysis of the gas sources in the fully mechanised working face 1001

For the coal seam 10# in the Wuhushan Coal Mine, the absolute gas emission quantity is 12.8m^3/min. When the fully mechanised working face 1001 is being exploited, the sources of the gas are shown in the Figure 1-23.

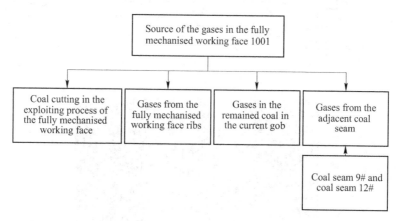

Figure 1-23 The gas sources of the fully mechanised working face 1001

B Comprehensive gas management technology of the fully mechanised working face 1001

Based on the problem that the gas may be over than the limitation which is resulted by the gas leakage problem when the fully mechanised working face 1001 is being exploited, the gas drilled boreholes at the upper position are prepared in the haulage roadway 1001 and the ventilation roadway 1001, to drain the gases. In the ventilation roadway 1001, a thin steel tube with a diameter of 400mm is set in the ventilation roadway 1001 is set, to drain the gas in the gob area and the upper corner. In the haulage roadway and the ventilation roadway, two sealing walls and corresponding local fans are installed, to implement the ventilation technology with uniform pressure. This successfully solves the problem of gas leaking in the fully mechanised working face 1001. Furthermore, this solves the problem of the gas overrunning in the fully mechanised working face under the combined condition of two layers of gob areas in the extremely adjacent coal seams. The gas drainage technology is shown in Figure 1-24.

Reasonable controlling of the wind direction in the gob area is acquired, as shown in Figure 1-25. Based on the continuous observation results of the gases and air volume of the fully mechanised working face, the intake air of the working face is ranged from 570 to 600m^3/min. As for the air volume in the ventilation roadway, it is ranged from 480 to 600m^3/min. In the production period of the working face, the gas concentration is ranged from 0.3% to 0.6%. The production of the working face has the tendency to be normal, realising the safety production.

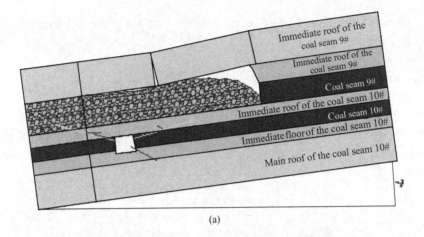

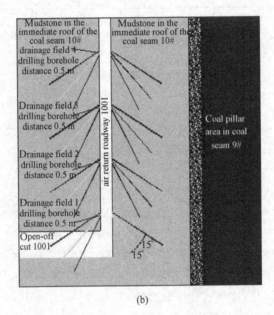

Figure 1-24 Layout diagram showing the drainage field in the ventilation roadway 1001
(a) Layout diagram along the dip direction of the drainage drilled boreholes in the ventilation roadway 1001;
(b) The vertical view showing the layout of the drainage drilled boreholes in the ventilation roadway 1001

The implementation of the exploiting and safety technology in the extremely adjacent coal seam successfully reduces the roof collapsing of the coal face of the fully mechanised working face and the accident of the coal rib spalling. This improves the safety of the fully mechanised working face. For the working face, the collapsed height is controlled around 0.5m. The maximum depth of the rib spalling is around 0.5m. Compared with the previous similar conditions, they are reduced by more than 75% and 50%. The comprehensive mechanised safety mining of the lower coal seam in extremely adjacent coal seams in the Wuhushan Coal Mine is first realised. This solves the connecting problems of the working face. Meanwhile, this improves the single production and the coal operator efficiency. Compared with the normal working face, the annual

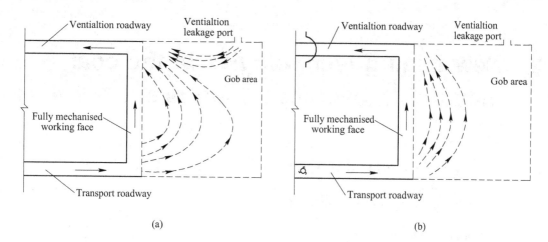

Figure 1-25 The schematic diagram regarding the air flowing direction vector in the gob area before the uniform pressure ventilation is conducted and after it is conducted

(a) Before the uniform pressure ventilation is implemented; (b) After the uniform pressure ventilation is implemented

production of the fully mechanised working face is doubled and the coal operator efficiency per person is improved by 3 to 4 times. This is beneficial for the concentrated production of the mine site and reducing the number of production mining areas. Overall, the coal exploiting level of the Wuhai Energy Corporation Limited is improved.

2 New mining technologies in the coal industry in overseas countries

2.1 Advances in mining and safety technologies in USA[5]

2.1.1 Safety technological developments

Safety performance is shown in Figure 2-1.

Figure 2-1 Safety performance

Major areas of safety:
(1) Methane control;
(2) Mine ventilation;
(3) Roof control;
(4) Electrical safety;
(5) Mine monitoring;
(6) Rock dust application.

The subjects chosen for discussion here are developments in methane control, roof control and respirable dust control.

2.1.1.1 Methane control

(1) Methane concentration control——1% and 2%.
(2) Air quantity and velocity.
(3) Methane and air measurements——foreman, machine operators, before shift, during shift.
(4) Machine-mounted methane monitors——continuous miners, longwall machines, cutting

machines, roof bolters, ——monitor to provide warning at 1% and cut off power at 2% on warning, operator to cut off power.

Atmosphericmonitoring system [AMS]. AMS consists of a system of hardware and software capable of (as shown in Figure 2-2):

(1) Measuring atmospheric parameters.

(2) Transmitting the measurements to a designated surface location.

(3) Providing alert and alarm signals.

(4) Processing and cataloguing atmospheric data.

(5) Providing reports.

Important parameters:

(1) Design of surface location and staffing.

(2) Location of AMS sensors—methane, carbon monoxide, belt airways, return airways, primary escapeway, electrical installations, diesel engine operating airways.

(3) Discriminating ability for fire, diesel.

(4) Actions in response to AMS alert, actions and malfunctions.

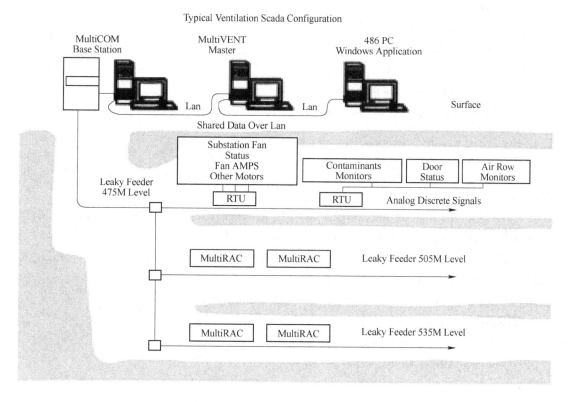

Figure 2-2 Typical mine monitoring and telemetry

2.1.1.2 Modern roof support technologies

(1) A temporary roof support system [ATRS]: device to provide temporary roof support from a

location where the equipment operator is protected from roof falls, as shown in Figure 2-3.

(2) Mobile roof supports [MRS]: device that uses shield technology and mounted on crawlers or tracks, as shown in Figure 2-4.

Figure 2-3　A temporary roof support system　　　　Figure 2-4　Mobile roof supports

MRS use and advantages:

(1) Full & partial extraction;

(2) Longwall shield recovery operations;

(3) Longwall head & tailgate applications;

(4) Removes miners from dangerous locations;

(5) Reduces roof support materials handling accidents and injuries.

2.1.1.3　Dust control methods

(1) Preferred order of control methods: 1) Engineering, 2) Administrative, and 3) Personal Protective Equipment.

(2) The Mine Act requires the provision of PPEs to miners where concentrations may be higher but prohibits the substitution of PPEs for engineering controls.

(3) The use of administrative control does not reduce the responsibility of the operator to maintain the dust level at or below required levels.

(4) In practice, all applicable methods are used as necessary.

Dust control in continuous mining (as shown in Figure 2-5～Figure 2-8):

(1) Continuous Miner Dust Control——Miner design;

(2) Continuous Miner Dust Control——Sprays;

(3) Spray Fan Application;

(4) Machine-mounted Scrubber;

(5) Modified Cutting Cycle;

(6) Ventilation Controls——Single split, double split and auxiliary ventilation schemes;

(7) Other Sources——Roof bolters, conveyors;

(8) Other Methods——Water infusion, foam application.

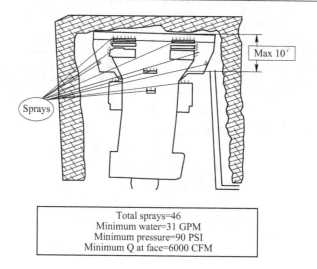

Figure 2-5 Spray system for a remotely controlled continuous miner with integral bolters

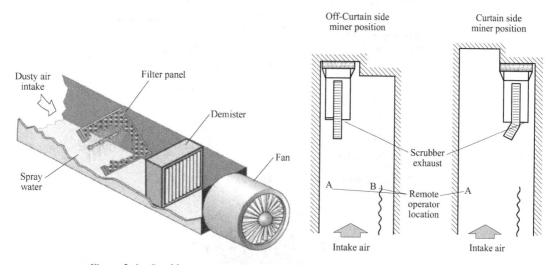

Figure 2-6 Scrubber unit

Figure 2-7 Operator positions for reduced dust exposure

Longwall dust control technology (as shown in Figure 2-9 and Figure 2-10):

(1) Shearer Dust Control——Shearer design;
(2) Shearer Dust Control——Shearer Sprays;
(3) Shield Dust Control;
(4) Stage Loader/Crusher Dust Control;
(5) Modified Cutting Sequence——Uni-Di;
(6) Ventilation Controls——Gob curtain;
(7) Ventilation Controls——Wing curtain;
(8) Remote Control and Automation;
(9) Other Methods——Water infusion, foam;

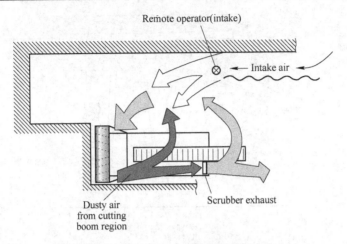

Figure 2-8　Remotely controlled scrubber miner

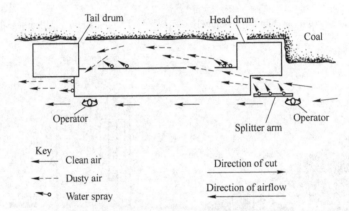

Figure 2-9　Shearer-clearer system operation

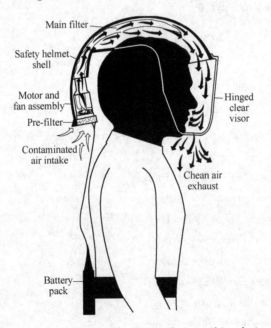

Figure 2-10　Helmet with a powered-air purifying device

(10) Other Sources of Control——Conveyors, reentrainment;

(11) Personal Protective Equipment——PAPR.

2.1.2 Mine safety legislative developments & safety organizations in the U.S.

2.1.2.1 Aspects of the current U.S. mine safety legislation

(1) The goal is to provide more effective means and measures for improving working conditions and practices. This goal establishes an overall structure to the legislation that is remedial in nature. While there are sections of the law that include enforcement and penalty provisions for violations, the general approach specifies accepted safety and health standards and procedures, with the understanding that most managers and workers will follow them in the interest of safety.

(2) Provisions of this legislation include:

1) Interim (temporary) safety and health standards;

2) Enforcement mechanisms (inspections, citations, orders);

3) Strict liability: Mine operators and managers are responsible for all actions and conditions. The only exception is smoking;

4) Penalties: Civil (monetary) fines are mandatory for safety and health violations. Criminal (personal) penalty provisions may be used in cases where individuals knowingly or willingly violated safety or health standards;

5) Appeals process and judicial review.

2.1.2.2 Safety organizations: the types and functions of safety organizations

Safety organizations exist to assure professional progress towards achieving greater health and safety for the workers. They can be of the following types:

A Governmental safety agencies

In the United States of America, the following federal agencies are primarily responsible for workplace safety and health:

(1) Mine Safety and Health Administration (MSHA)——enforcement of the 1977 Mine Health and Safety Act, and mine safety regulations (CFR Title 30).

(2) Occupational Safety and Health Administration (OSHA) and Environmental Protection Agency [EPA].

(3) At the state level, there can be agencies similar to MSHA and OSHA with health and safety responsibilities for workers in their state. For example, the Bureau of Deep Mine Safety in Pennsylvania is one of the oldest and active health and safety agencies for miners.

(4) National Institute for Occupational Safety and Health (NIOSH) through their mining laboratories in Pittsburgh and Spokane. The results of the research (in the areas of ventilation and dust control, methane control, safe use of explosives, and safe use of electricity) contribute extensively to improved, safer design of mines.

B Industrial safety organizations

(1) These associations address industry wide health and safety concerns.

(2) They serve as a resource for their clients, the general public and the government.

(3) In the mining industry, there is a national trade association with one of its major functions defined to support its members with health and safety issues.

(4) Such associations also exist at state levels.

The Joseph A. Holmes Safety Association (HSA) was formed by 24 leading national organizations in 1916 with the objective to prevent fatalities and injuries and to improve the overall health among officials and employees in all phases of mining. The association operates through 4415 local chapters throughout the United States, and its membership includes government, mining companies and suppliers. This volunteer organization exists due to the active support of health and safety professionals and interested miners in meetings, and in sharing emerging problems and workable solutions. There is an annual meeting of the HAS; there are also several meetings of the chapters. HSA publishes timely and useful materials and makes much coveted awards and scholarships for safety leadership to encourage new generations to enter the health and safety professions.

C Professional and technical societies

(1) Health and safetyis a distinct profession.

(2) Several societies address special needs of the health and safety professionals: e. g., American Society of Safety Engineers (ASSE) with its many divisions; American Industrial Hygiene Association (AIHA).

(3) Most societies publish magazines and hold annual meetings. These provide excellent forums for safety professionals to share information, knowledge and experiences.

(4) Some professional societies certify their members, for careers in health and safety, through competency examinations. For certain jobs, certifications may be required.

2.2 Borehole mining[6]

Borehole mining (BHM) is a remotely operated method of extracting rock material through boreholes by high-pressure water jets. Figure 2-11 illustrates a BHM tool and fluids circulation. The tool consists off at least two pipes, one for pumping down high-pressure water, the second for delivering slurry to the surface. This section reviews the BHM process. It discusses current and new applications for borehole mining and help the readers determine if BHM is applicable in their situations.

2.2.1 How BHM works

In BHM, a borehole is drilled to the desired depth (production interval). A casing column is then lowered down the hole. Since BHM takes place in an open hole, the casing shoe is located just above the production interval (orebody). The borehole mining tool is lowered into the hole.

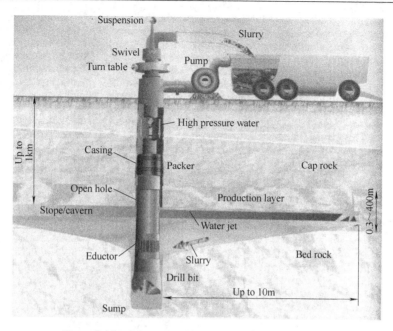

Figure 2-11 Schematic of borehole mining tool operation

The tool usually has:

(1) A drill bit to re-drill collapsed intervals.

(2) The eductor section to pump the slurry to the surface.

(3) The hydro monitor section, containing the cutting nozzle.

(4) An extension section.

(5) A hub, connecting it all to a drill pipe string. This string is extended up to the land surface.

On the surface, the tool has a swivel. It allows rotation in the hole and connects to a water supply and a slurry collector. The tool is lowered until the cutting nozzle reaches the required depth. Then high-pressure water pumping is started. Table 2-1 shows some of the technical data associated with borehole mining.

Table 2-1 Borehole mining technical data

Parameter name	Parameter index
Water pressure required	70~200atm (1000~3000psi)
Water flow rate required	150~300m³/h (500~1000gpm)
Productivity	Up to 90m³/h (3000 cu ft/hr)
Slurry consistency (rock/water)	1/10~1/1
Depth of mining	20m~1km (60~3300ft)
Production interval	0.3~400m (1~1300ft)
Tool outside diameter	89~305mm (3.5~12in)

As BHM progresses, underground cavities or caverns are created. The tool can be rotated or moved up and down along the borehole, or in any combination, creating various shapes of caverns (Figure 2-12).

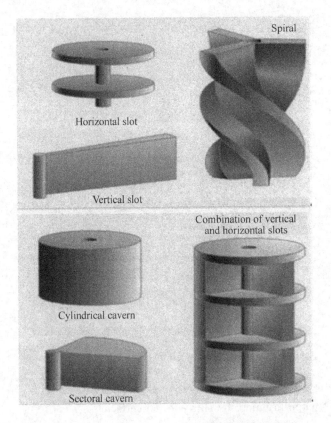

Figure 2-12　Possible cavern shapes with borehole mining

2.2.2　Current BHM applications

BHM has been used to extract several types of mineral resources. Some them include phosphate, uranium, iron ores, coal, bauxite, gold, diamonds, quartz sends, fine gravel, titanium, zirconium, germanium, other mineral resources and industrial materials. However, the application of borehole mining goes beyond just mining. It includes exploration, solving of environmental problems, underground construction and more. Table 2-2 illustrates some of the versatility of BHM.

The versatility of BHM is further exhibited as these applications are executable from land surface, openpit floors, underground mines or water surfaces as the circumstances require. Figure 2-13 shows the BHM tool being lowered into a coalbed methane well where it was used to stimulate gas production.

Figure 2-13　Borehole mining tool going into a borehole

2.2 Borehole mining

Table 2-2 Primary borehole mining application areas

Application	Purpose	Description
Exploration	Bulk sampling	Extracting of large volume of material through single hole instead of drilling numerous cores
Mining	Minerals and industrial materials	Extracting of materials through boreholes pre-drilled from land or water surface, openpit floor or underground mines. Under-reaming of the well within the production interval
Stimulation	Oil, gas, water In-situ leaching Solution mining	Increase orebody permeability and control of reagent flow by creating vertical slots and caverns into the orebody. Increasing water/mineral contact surface by pre-cutting vertical slots on the first stage of solution mining
Construction	Foundations Subsurface storage	Vertical slots or cylindrical caverns filled by concrete. Caverns for storage of oil and gas or for waste material disposal
Environment	Waste collectors, Underground walls Acid drainage in-situ treatment	Extended vertical slots to collect the contaminated aquifer. Vertical slots filled with clay/mud or concrete. Injection of limestone slurry into abandoned, inaccessible mines

2.2.3 New BHM applications

Much BHM experience has been gained throughout the years in different countries and different applications. But there remains one disadvantage to borehole mining that has slowed its industrywide adoption. The ratio of drilling and completing a borehole versus the cost of the actual borehole mining may vary from 0.5 to 10 or higher.

An example would be a block of ore 50m×50m×3m (150ft×150ft×10ft) and 100m (300ft) deep. This block would require about 23 vertical drill holes (assuming a 4m or 13ft cutting radius and 2m or 6ft pillar between caverns). To drill and case one of these boreholes may be several times more expensive than to extract the ore. For stimulation, construction, environmental and other applications, this may not be as important as for mining. In this example, borehole mining through a horizontal borehole driven along the production layer would significantly improve the overall drilling/mining ratio and further enhance BHM technology.

In the past decades, there were several attempts to turn BHM 90°. These ideas, however, suffered from a complicated technology or expensive tooling.

In 2003, a patent-pending status was granted to the horizontal borehole mining technology that could solve that long-time problem and further increase BHM versatility. The invention uses a collapse of the rock after it is undercut. This collapse occurs when the span between pillars reaches a critical distance. So, if the waterjet can create an extended horizontal cut, the rock will collapse under pressure.

This horizontal undercut. or slot, is easy to create through a horizontal borehole. The BHM tool is placed so the hydromonitor cutting nozzle is oriented near-horizontally. The tool is then moved along the borehole without rotation so that the jet creates an extended undercut. The rock collapses to the slot where the water jet loosens and breaks the rock, creating a pregnant slurry. The eductor pumps it away from the borehole.

Using this invention, borehole mining is applied through near-horizontal boreholes driven from an openpit floor or an underground mine. It also is applicable through a motherwell that is pre-drilled from a land or water surface first vertically and then deviated to the required direction. In the above example, a single vertical well with 10 deviated/horizontal legs could mine the same ore block, but with significantly less drilling costs and increased recovery. Figure 2-14 compares the sample ore block mined by vertical and horizontal BHM techniques.

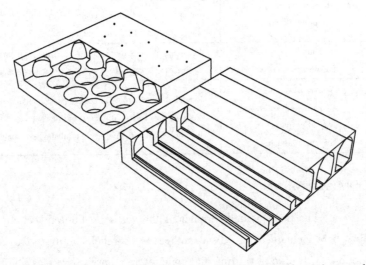

Figure 2-14　Comparison of borehole mining with vertical holes and with horizontal holes

2.2.4　When to use BHM

To adopt a borehole-mining concept, it is important to understand that BHM is not a panacea. The effectiveness of BHM in any particular case may not compete with conventional mining methods. It depends on several parameters, including the price of the mining mineral, ore physical properties, hydrogeology and many others.

In most cases, borehole mining cannot compete with conventional methods. But there are circumstances when BHM may have no alternatives, and can therefore supplement conventional mining. In general, the more difficult the mining/operating conditions, the more likely borehole mining will be advantageous.

Borehole mining should be considered in areas with thin production layers, low ore concentrations, difficult hydrogeological situations, hazardous conditions, underwater/offshore operations and inaccessible locations. Although incomplete, this list provides some situations when borehole mining may compete with traditional technologies or even have no alternatives. Water is recycled while borehole mining and it is not very difficult to warm it slightly to avoid freezing. So desert and polar zones could be added to the list. Table 2-3 lists some advantages to borehole mining.

Table 2-3 Borehole mining advantages

Advantage	Description
Safety	No personnel underground, all mining performed remotely by drill rig and pumping crew
Ability to work in remote areas	No need for electricity or other infrastructure. If hole can be drilled, borehole mining can be performed
Minimal environmental impact	Small footprint of disturbance, water is working agent and is recycled
Mobility	Limited equipment (primarily a drill and pumps) easy to relocate
Selectivity	Can mine only ore zone and control shape of cavern (see Figure 2-12)
Wide area of application	See examples in Table 2-2
Automated & computerized control	Easy to install sensors and control process
Low capital & operating costs	Small amount of equipment can be readily rented or purchased, limited labor, horizontal BHM lessens drilling
Simplicity	Pump water down, bring slurry up
Ability to work in range of conditions	Underwater, hazardous/explosive/flammable circumstances, in or under unstable or collapsing intervals

2.3 The mining technology in open-pit mining[7]

The open-pit mining indicates disclosing and exploiting coal resources or other mineral products directly from the ground. Compared with the underground mining, the open-pit mining has many advantages. For example, the mine site construction period is short. It is easier for the working face to be launched. Additionally, the production ability is large and the labour production rate is high. The resource recovery rate is high while the production cost is low. Furthermore, the safety condition is superior and the environment is fine. Therefore, if the natural condition is permitted, the open-pit mining method should be considered preferentially. With the development of the global economy, the demanding on coal resources is increasing dramatically. The percentage that the quantity of coal resources exploited with the open-pit mining accounts in the total exploited coal resources in the world is gradually increasing. When the open-pit coal mining is being developed rapidly, the countries have innovations in the aspects of the safety production and the advanced large-scale equipment. The miracles and records in the coal development history have been created, which has already become the star of the world industry progress.

2.3.1 Classification of the open-pit mining methods

In recent years, the mechanised mining method is regarded as the main open-pit mining method. Based on the reserve status of the mineral resources, different countries use different open-pit mining methods. The most commonly used open-pit mining methods include: mining method with pitting, overcasting mining method, quarrying mining method, auger drilling mining method and mining method with water.

2.3.2 Technical procedures in the open-pit mining method

2.3.2.1 Technical procedures in production

In the normal stripping and mining production process, the open-pit mining method generally includes the following technical procedures, including preparation on the ground surface, borehole drilling, blasting, tunneling and exploiting, transport and land reclamation.

2.3.2.2 Auxiliary technical procedures

For the open-pit mines, to maintain the normal production, besides the above-mentioned production procedures, the auxiliary technical procedures are also needed, including the power supply (such as the electricity supply and the petroleum supply), the equipment maintaining, water drainage and unwatering, maintaining of the slope and communication.

2.3.3 New technologies and new equipment in overseas open-pit mining

2.3.3.1 Drilling machine

The 49R drilling machine invented by the Bucyrus International Company of the United States of America has a number of characters, such as the rack and gear pressure device without chains, the new driving system, the planetary transmission state pressure hydraulic propulsion device without chains (which can reverse the rotating direction). The Bucyrus International Company further tries to improve the operation safety and the degree of comfort for drivers and maintenance crew, through adopting the operators' cab which has the function of noise reduction based on increasing pressure and sealing and temperature controlling, together with the automatic levelling system based on four hydraulic lifting jacks.

The mine site used drilling machine P&H250XP manufactured by the R&H Company has already entered the Venetia diamond deposit mine belonging to the Anglo-American Company which is located in the Northern Province of South Africa. In the Venetia diamond deposit mine, the diesel 250XP-ST drilling machine with the high drilling rig is used. In the strong rock masses, the drilled borehole has a diameter of 251mm and the depth of the borehole is up to 20m.

The Reedrill drilling machine company a new SD-250 crawler-type drilling jumbo driven with the hydraulic pressure. Additionally, the HPR-45 rock drill driven with the hydraulic pressure is matched. The diameter of the drilled borehole ranges from 64mm to 89mm. The electric power is supplied with a diesel generator with the capacity of 130kW. The fuel consumption is 23L/h. This equipment can work continuously for more than 13 hours and it is not necessary to refuel in this period. The gas supply capacity of the air compressor is $250ft^3/min$ ($1ft^3/min = 0.0283m^3/min$). The cooler is installed side by side, which is beneficial for improving the cooling efficiency and convenient for cleaning.

2.3.3.2 Loading equipment

A The electric shovel

At present, the oil sand deposit mine in the Province of Alberta in Canada has already become the principal market of the large scale electric shovel. In November, 2001, the first introduced P&H 4100BOSS electric shovel has already started the spading operation in the Aurora Mine which belongs to the Syncrude Company. The BOSS electric shovel is the upgrading product of the 4100TS electric shovel which has extremely favourable application effect under the condition of low soil bearing capacity. The BOSS electric shovel accelerates the velocity of the spading cycle, relying on increasing he lifting power of the bucket and the declining power of the bucket. The advanced computer controlling system is adopted for these operation functions.

Figure 2-15 indicates the P&H BOSS electric shovel used in the oil sand deposit mine belonging to the Syncrude Company. The Optima bucket is matched. Specifically, each bucket can spade the ground surface soil up to 110t.

Figure 2-15 P&H BOSS electric shovel

B Hydraulic pressure shovel

At the end of 2000, the Caterpillar Company formally introduced the 5230B working face hydraulic pressure shovel which is especially developed for the large scale open-pit mine and the realistic associated equipment for the earthwork hydraulic pressure shovel Cat785B vehicle. At the same time, it can operate and work efficiently with the Cat/789B vehicle and the 793C vehicle. The 5230B hydraulic pressure shovel has two different matched modes. One is that for the mining and excavating working face with the bucket volume of $17m^3$, the frontal shovel is used. Specifically, under this condition, the spading load is 1162kN and the pushing load is 1145kN. On the other hand, for the earthwork operation with the bucket volume of $16m^3$, the reverse shovel is used. Under this condition, the spading load is 855kN and the load in the rod is 885kN. This hydraulic pressure shovel is designed based on the requirement that the filling coefficient of the shovel is high and the cycling time in spading and excavating is short. The hydraulic pressure shovel is driven by a Cat3516B EUI motor. The net power of the motor is 1156kW and the rotating speed is

1800r/min. The working weight is 328.1t. The 5230B electric shovel is the biggest in the series of electric shovels in this company. Compared with the 5230 electric shovel which has been replaced with the 5230B electric shovel, its motor power increases by 5%. Additionally, the hydraulic flowing quantity increases by 10%. To increase the durability and reliability of the 5230 hydraulic pressure shovel which are developed in 1994, the Caterpillar Company improves the equipment structure and the operating system. This makes the mass of this equipment increase 13t.

2.3.3.3 Transporting equipment

A Vehicles used in mine sites

Only relying on larger scale vehicles can decrease the mining cost. The design of mine sites and the infrastructure must have some special modification and then can adapt to large scale vehicles. Different kinds of excavators which can load materials for more larger scale vehicles have already been equipped. However, usually, the required vehicle accessories, tyres and motors should be developed in advance. Then, the vehicle manufacturer can start designing more larger scale vehicles. Nevertheless, if the specification of the vehicle is larger, the resulted production loss is higher on the condition that malfunctions occur. Therefore, to decrease the transporting cost, it is required that management should be carried out on the dead weight capacity, the lifetime of the accessories, the cost and the efficiency. In fact, the reliability and the serviceability ratio of the vehicle are the core.

B Articulated dump truck vehicle

In the mine sites where severe condition may occur on the road surface, the articulated dump truck (ADT) can be adopted. This kind of truck has obvious superiority when it is used to transport smaller load. In June 2015, the Komatsu Company the 40t HM400-1 articulated dump truck which has a maximum weight up to 66666kg. This vehicle has an effective load carrying capacity of 40t when it is moving with the maximum nominal speed of 60km/h. The proportion between the effective capacity and the total weight is excellent. Specifically, more than 55% of the total weight is the effective capacity. Therefore, the mine site can transport more materials. Furthermore, it is not necessary to carry dead load. It also has many other advantages. For example, more fuel can be saved. Additionally, the operation cost of the company can be decreased. Thirdly, the gradeability or the climbing ability of the vehicle is increased. Last but not least, the downgrade speed is accelerated.

The HM400-1 articulated dump truck is composed of 6 wheels. For the braking system, the continuous oil cooled multi-disc method is used, which has excellent braking performance. Additionally, this system is completely sealed. As a consequence, the invading of the dirt can be prevented. Therefore, it has much higher reliability and the corresponding lifetime is much longer. For the HM400-1 type, the continuous oil cooled multi-disc speed reducer is adopted. It is not necessary to conduct the frequent operation and then the speed reducer can work and decelerate.

Therefore, the vehicle can advance safely with a high speed. It should be mentioned that the vehicle can realise this even if it is advancing on the slope which is not only long but also steep. Additionally, this articulated dump truck with the 6 × 6 wheel driven mode is equipped with the front axle locking convertor and the rear axle locking convertor which can be operated conveniently. Furthermore, on the 3 axles, the differential locking device is installed. They can effectively drive under any circumstance.

The Volvo Company introduced the A25D type and the A30D type 6×6 articulated dump truck. The carriage volume of those two types of vehicles are $15m^3$ and $17.5m^3$ respectively. The load carrying capacity of them is 24t and 28t respectively. They are matched with the A35D type with the load carrying capacity of 32.5t and the A40D type with the load carrying capacity of 37t, forming a series. The characters of the D series articulated dump truck is that many different new technologies are used. Through increasing the load carrying capacity, accelerating the uninstalling speed, improving the braking system and extending the maintenance period, the transport efficiency can be improved. At the same time, the safety level is increased. On the other hand, the operation cost is decreased. Additionally, the extent of the environment pollution is decreased.

The Beier Company developed the D series new articulated dump truck which includes four types, namely B25D, B30D, B35D and B40D. The effective load carrying capacity is ranged between 25t and 40t. Its transport efficiency is much higher. Additionally, more fuel can be saved. Thirdly, the operation cycle is faster. Last but not least, it is more convenient for drivers to operate.

2.3.3.4 Monitoring of the equipment

The GPS (Global Position System) provides an increasing number of benefits for the most advanced open-pit mines in the world in terms of improving the production efficiency. Each mine site is seeking the dynamic monitoring system which is completely applicable for data collecting, mine planning, mine observation, dynamic displaying of the equipment maintenance and the overall management of the equipment. After this system is linked with the GPS, they can generate practical and beneficial consequence. The comprehensive mining information system MineStar which is jointly developed by the Caterpillar Company and its allied parter, is one of this kind of systems that can be supplied in this current stage. The MineStar system includes the monitoring of the equipment condition and the operation efficiency, tracking of the equipment and material flowing, management of the borehole drilling machine, CAES (Computer Aided Earthwork System) and the advanced vehicle planning and dispatching program.

The MineStar system has the ability in linking the equipment located in the field with each MineStar office system. Additionally, the MineStar system has the ability in linking other mining information systems.

2.3.4 Development tendency of the open-pit coal mining

(1) The proportion of the coal resources exploited with the open-pit mining will still increase

stably in the world.

In the long term, the proportion of the production output of the coal resources exploited from the open-pit coal mines in the total coal production of the world will still increase in the future. This is because the open-pit coal mining has a number of advantages. For example, the production ability of the open-pit coal mining is large. Additionally, the construction period for the open-pit coal mines is short. Furthermore, the labour productivity ratio of the open-pit mines is higher. Fourthly, the investment per unit ton coal is relatively smaller. And with the open-pit coal mine, the resource recovery ratio is higher. Sixthly, the labour operation condition is better. Additionally, the open-pit coal mining is beneficial for safety production. Last but not least, there are a number of coal basins which are applicable for open-pit coal mining in each principal coal mining countries. Under the environment that the energy requirement of the world is continuously increasing, the open-pit coal mining industry of the world is still confronted with a vast quantity of development opportunities.

(2) The open-pit coal mining and production have the further tendency to become large scale and intensification.

From the perspective of the whole world, intensification exploiting of the coal resources will become the main tendency. One is that the principal open-pit coal mining countries in the world are continuously increasing the production level in terms of intensification and large scale in the future. Through the intensification production, the resource recovery ratio can be increased. Additionally, the economic cost can be decreased dramatically. Secondly, the powerful large scale coal companies develop, uniformly becoming the transnational giant energy group. Then, the intensification and high efficiency can be continuously conducted in terms of the operating management of the companies, the acquiring of the resources, the processing of the resources and the selling of the resources.

(3) The technical and technology innovation based on the mining mode.

With the gradual transferring of the open-pit coal mines from the previous single extensive exploiting mode in which on the production capacity and the production output are emphasised, to the multiple development objectives including the large scale, the intensification and the green ecology, the open-pit coal mining technology and technique will continuously rely on the development of the modern science and technology. Furthermore, the driving force of the interdisciplinary science will also be relied on. Then, it will seek the new round of innovating and upgrading, trying to meet the requirement of new exploiting mode.

(4) On the basement of large scale and forming a complete set, the equipment will further seek more intelligence.

With the continuous development of the techniques and technologies in the open-pit coal mining, the requirement of the corresponding technical parameters on the equipment is becoming higher and higher. This will lead to the results that the equipment will be continuously improved to meet the technical requirement. In particularly, it will try to meet the requirement that extremely large scale coal mines will conduct the resource exploiting operation under the severe environment.

Additionally, it will also try to meet the requirement of pursing and seeking on low costs and high profits. The exploiting equipment will definitely develop towards the direction of intelligence based on the large scale and the complete set.

(5) The systematic development of the green exploiting technology in the open-pit coal mines.

In the whole world area, the construction level of the green mine sites is continuously increasing. A series of environmental problems such as the destruction of the ground surface and the slippage of the slope which are resulted by the open-pit coal mining, compel the open-pit coal mines to construct the systematic green exploiting technical system. During this process, the open-pit coal mines will take the basement of the integration of the current exploiting and reclamation, energy conservation and environment protection, and developing the recycling economy. Additionally, the latest outcomes in the corresponding areas such as the techniques, the technologies, the equipment and the management will be relied on. In this way, the green exploiting technology of the open-pit coal mines will be further developed and completed.

(6) The through development of the digitalisation and intelligence of the mine sites, realising the smartness in the end.

The digitalisation of the mine sites is beneficial for improving the production efficiency dramatically. It is also the significant guarantee for the mine sites to realise the dynamic monitoring and safe management. The final objective of the digital mine sites is to realise the intelligent mine sites in the end. During this process, several basic modules will be regarded as the basement, such as the data collecting system, the model of the digital mine sites and the dynamic monitoring. Additionally, the development of the internet of things technology will be used for reference. Then, the intelligence level of the mine site exploiting will be continuously increased. Finally, the objective of intelligent mine site will be realised.

3 Research progress on the ground pressure in the longwall face and its controlling technology

3.1 Forecasting and the research progress of the mine roof disasters

In China, the number of deaths that are resulted by the mine roof disasters is in the first place in kinds of disasters. Among them, the number of deaths resulted by the coal mine roof disasters accounts for 76.1% and the number of deaths resulted by the non-coal mine roof disasters account for 23.9%. In the total number of deaths resulted by kinds of disasters in coal mines, the roof disasters account for nearly 40% of all disasters. For the coal mine roof disasters, the disasters resulted by the local fractured instability of the roof rock masses account for 81%. The disasters resulted by the rupture and instability of the large scale strong roof account for 13%. The disasters resulted by the rupture instability impact of the large scale strong roof account for 6%. Figure 3-1 and Figure 3-2 show the average annual number of deaths in kinds of mine disasters in recent years and the percentage of the deaths resulted by kinds of roof disasters respectively.

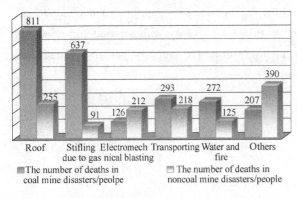

Figure 3-1 The average annual number of deaths in kinds of mine disasters in recent years

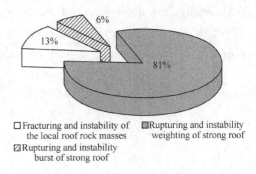

Figure 3-2 Percentage of the deaths resulted by kinds of roof disasters

3.1.1 Mine roof disaster state and the rock mass environment

According to the geological condition and the roof collapsing state, the roof rock masses and its disasters can be classified as four types: (1) large area hanging roof rock masses and the ruptured roof collapsing, (2) Blocky fractured structure rock masses and slippage roof collapsing, (3) fractured structure rock masses and the loose leakage collapsing, and (4) thin layer composite rock masses and separated thin layer collapsing. The roof rock masses of four types and its corresponding disasters are shown in Figure 3-3.

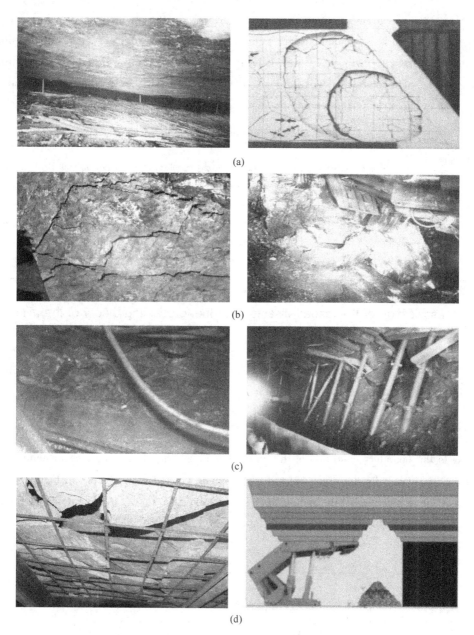

Figure 3-3 Classification of the roof rock masses and its disaster
(a) Large area hanging roof rock masses and the ruptured roof collapsing;
(b) Blocky fractured structure rock masses and slippage roof collapsing;
(c) Fractured structure rock masses and the loose leakage collapsing;
(d) Thin layer composite rock masses and separated thin layer collapsing

The immediate roof of the fully mechanised working face is located above the coal body that is to be exploited and the hydraulic supports. However, the section where the local engineering disaster occurs is mainly located at the coal face of the working face and the transition area of the hydraulic supports. The influence parameters of the stability of rock masses is mainly influenced by its self-characters and the environment where it is located. However, the environment that influences the stability of rock masses, include the natural geological environment and the production environment. Therefore, in this area, the stability of the immediate roof is influenced by kinds of parameters. The parameters that will influence the mine longwall face roof disasters are divided into five types. The specific influencing parameters are tabulated in Table 3-1.

Table 3-1 Influencing parameters of the mine roof disasters

Main types of the roof rock mass environment	Mechanical characters of rock masses			Characters of beddings and fractures		Geological structures	Mining and tunneling conditions		Main roof conditions
Influencing parameters	Tensile strength /MPa	Cohesion /MPa	Internal friction angle /(°)	Thickness of layers /m	Spacing between joints/m	Lateral coefficient (λ)	Resistance force coefficient /%	Distance to the coal face /m	Position of the rupture line/m

3.1.2 Classification of the leaded disaster and the forecasting theory of the mine roof disasters

3.1.2.1 Mechanism and the forecasting action of the large area rupture and instability of the mine roof

A Mechanical model of the large area roof rupture

According to the boundary condition of the mine exploiting, the plate structure mechanical model of the elastic basement for the mine roof large area is constructed[8], as shown in Figure 3-4. The partial differential equation with high orders is solved. And then, the main stress field and displacement field of the plate structure can be acquired, as shown in Figure 3-5.

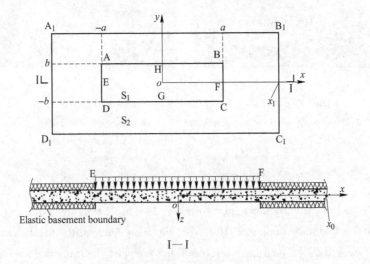

Figure 3-4 The main roof thin plate model with the elastic basement

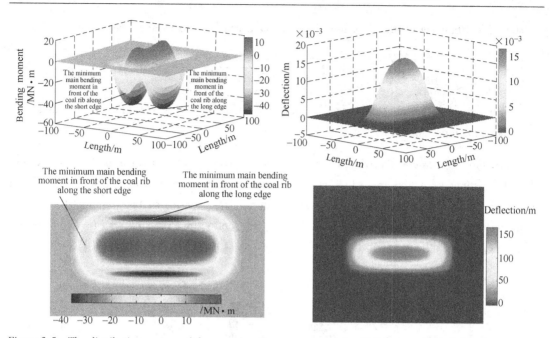

Figure 3-5 The distribution contour of the main bending moment and the displacement field of the plate structure

B Instability criteria of the large area rupture and the engineering disaster condition

The main roof generates arc-shaped triangle rupture block in the end region. In the middle area of the working face, the shape of the main roof that is ruptured approximately has the trapezoid geometry. Based on this, the plate-type brickwork structure model is established. As shown in the Figure 3-6, along the advancing direction of the working face, the rock block A is the main roof that is above the supporting area of the coal face. The rock block B is the main roof that has already ruptured behind the rock block A. The rock block C is the main roof that is above the re-compressing area which is behind the rock block B. The intersection line of the rock blocks or rock plates A and B is the rupture line of the main roof that is above the working face. Along the direction that is perpendicular with the advancing direction of the working face, there are also similar three types of rock blocks or rock plates A, B and C.

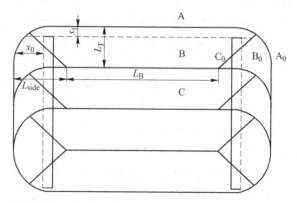

Figure 3-6 The schematic diagram of the rupture form of the main roof

The critical balance condition of the trapezoid plate when it rotates and loses its stability is:

$$\frac{L_T Q_0}{(h - L_T \sin\theta_2)^2} + \frac{4L_2 Q_1 \cos\theta}{(2h - L_2 \sin\theta_1)(h - L_T \sin\theta_2)} \leq 0.3\sigma_c \tag{3-1}$$

C The relationship between the extending progress of the large area rock plate rupture and the rebound compression field

In the Figure 3-7, around the rupture line area, the rebound field is defined as the rebound area I. The C-shaped rebound area which is in front of the rupture line is defined as the rebound area II. In the extending progress of the large area rupture of the rock plate, the form and the partition characteristics of the rebound compression field is: the semi ellipse rebound area I → compression area → C shaped rebound area II.

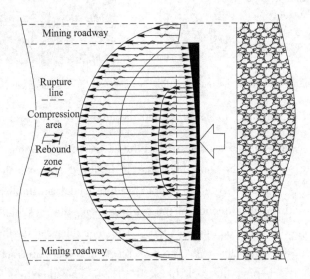

Figure 3-7 The rebound compression field of large area rupture of the rock plate

The advancing rupture occurs in the main roof. The rebound compression phenomenon immediately occurs in front of the rupture line when the rupture happens. Additionally, when the rupture occurs, due to the fact that under the main roof, it is supported by the immediate roof and the coal body, therefore no apparent weighting occurs. At this time, the main roof can reach the temporary stable state. Until the working face advances to the area of the rupture line, apparent weighting of the main roof occurs. It can be seen that rupture of the main roof and the phenomenon of the rebound compression occurs at the same time. Furthermore, the working face keeps advancing to the area of the rupture line, then apparent roof weighting occurs. It means that the apparent weighting of the working face lags behind the rupture of the main roof. Consequently, relied on this time lag, the pressure recording instruments can be installed in two roadways to measure and monitor the large area rupture of the coal face that is in front of the rock plate in advance. Furthermore, early warning can be conducted.

D Physical geography information induced by the rupture of the mine site roof plate

(1) The rebound compression information induced by the rupture of the large area plate structure. The large area rock plate rupture experiments draw the conclusion that rebound compression information can be generated by the rupture of the rock plate. Rebounding and compression points occur in sequence in the area that is in front of the rupture line and the area of two roadways. Furthermore, the maximum rebounding occurs in the middle of the working face while the minimum occurs in the area of two roadways. The rebound compression field character when the rock roof collapses with large area is shown in Figure 3-8.

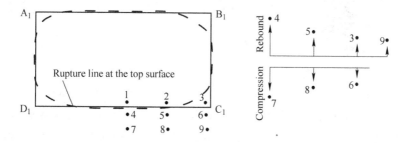

Figure 3-8 The characteristics of the rebound compression field when the large area rupture of the rock plates occurs

(2) The information of the acoustic emission induced by the rupture of the large area plate structure.

When the large area plate structure starts rupturing, a quantity of acoustic emission phenomenon occurs around the rupture line area of the long edge and the short edge. The point source of the acoustic emission has the "O" shape. In the middle of the gob area, it is also the concentration area of the acoustic emission point sources. They are mainly distributed in the area of the middle rupture line. The point source of the gob area has the "X" shape. The acoustic emission point source in the area of the coal pillar is mainly located in the middle area of the coal pillar and close to the working face side. The number of acoustic emission point source is 4~6 times of the normal state when the large area plate structure initially ruptures.

3.1.2.2 Disaster and forecasting theory of fracture and instability of the local mine roof rock masses

A Mechanism and disaster condition of the slippage roof collapsing of the roof fractured structure rock masses

The mechanical model of the wedge block and corner block when the core block of mine roof blocky structure rock masses slips is constructed. The distinguishing equation regarding the critical balance state of the core block is deduced and constructed. The core block slippage mechanical model and the distinguishing equation in the balance state are shown in Figure 3-9.

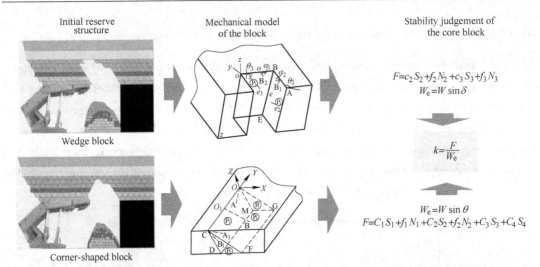

Figure 3-9 Mechanical model of the core block slippage and the distinguishing of the balance state

Analysing the relationship between the slippage instability of the core block and its influencing parameters, it can be acquired that the core of leading to the slippage instability disaster is that the anti-slippage force on the fractured surface decreases, which is resulted by the mining. Then, the critical condition that the hydraulic supports maintain the stability of the core block can be acquired.

The minimum support force of the hydraulic supports to keep the core block of the roof maintaining stable can be expressed as:

$$T = \frac{W_e - F}{2f_2 \sin\alpha_2} \tag{3-2}$$

B Mechanism and disaster condition of the loose leakage and collapsing of the roof fractured structure rock masses

Through the leaking and collapsing state of the fractured roof rock masses which is shown in Figure 3-10, the mechanical loose leakage model of the mine roof fractured structure rock masses can be constructed, as shown in Figure 3-11. Then, the definite relationship between the collapsing height and the supporting force can be acquired.

Figure 3-10 Leaking and collapsing state of the fractured roof rock masses

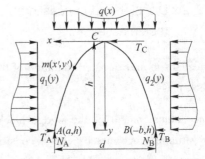

Figure 3-11 The mechanical collapsed arch model of the fractured roof

When the horizontal supporting force of hydraulic supports points at the coal face direction, the collapsed height is h_1 and when it points at the gob area direction, the collapsed height is h_2.

When the horizontal supporting force of hydraulic supports (T_A) points at the coal face direction:

$$h_1 = \sqrt{\left(\frac{T_A}{\lambda q}\right)^2 + \frac{d^2}{4\lambda}} - \frac{T_A}{\lambda q} \tag{3-3}$$

When the horizontal supporting force of hydraulic supports points at the gob area direction:

$$h_1 = \sqrt{\left(\frac{T_A}{\lambda q}\right)^2 + \frac{d^2}{4\lambda}} + \frac{T_A}{\lambda q} \tag{3-4}$$

C Separated thin layer collapsing mechanism and disaster condition of the thin composite roof rock masses

In the mine longwall face, for the bedded structure roof which only has horizontal beddings and has a small number of joints, the local instability and the collapsing are mainly resulted by the fact that after the roof is subjected to the vertical stress of the overlying rock strata, bending and sinking occur. However, for the rock materials, they can be regarded as the elastic-brittle material. In the bending and sinking process, the rock elastic modulus in different layers determine that different sinking extents can occur. Consequently, the difference on sinking can occur. This sinking difference is the normally expressed bed separation.

Assuming that in the roof of the fully mechanised working face, towards the up direction, the elastic modulus and the thickness of the first, second and third layers are E_1, E_2, E_3 and h_1, h_2, h_3. Then, the essential condition to generate bed separation is:

$$3E_1 h_3^2 \sum_{i=2}^{3} E_i h_i^2 > 2E_1 h_3^2 \sum_{i=1}^{3} E_i h_i^2 > \sum_{i=1}^{3} E_i h_i^2 \sum_{i=2}^{3} E_i h_i^2 \tag{3-5}$$

In the bed separation process of the longwall face roof, the roof sinking increases continuously. Meanwhile, the roof beam model under the support condition of the hydraulic supports and coal face is constructed, as shown in Figure 3-12.

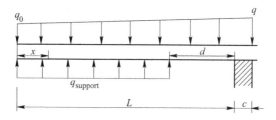

Figure 3-12 The mechanical model of the separated thin layer collapsing

Among them, the length of the lowest immediate roof simply supported bean is L; The supporting length of the hydraulic supports is $L_{support}$; The distance of the empty roof in the cross-section is d; The supporting strength of hydraulic supports is $q_{support}$; The above load is the trapezoid load and its strength ranges from q_0 to q; The thickness of the roof is h.

The rupture condition when the lower surface of the immediate roof is subjected to tensile load is:

$$\sigma(x = x|_{M(x) = M_{\max}}) > \sigma_s \qquad (3\text{-}6)$$

Assuming that the roof thickness is the length of a hydraulic support, namely b. And the height is h. The shear strength is τ. Then, the supporting point of the roof coal face, B, can provide the maximum shear force, which is τbh. It is assumed that at the point B, due to the loss of the roof integrity which is resulted by the tensile failure, the rock mass tension coefficient is ξ. Then, when the rock mass does not have tension crack, $\xi = 1$. When the rock mass has the full tension crack, $\xi = 0$. Therefore, practically, the maximum shear force that the point of B can provide is $\xi g \tau bh$.

Then, the tensile and shearing rupture disaster condition is:

$$Q_B = q_0 L_s + \frac{q_0 L}{10} - q_s L_s > \tau \xi bh \qquad (3\text{-}7)$$

Q_B is the shear force at the roof coal face supporting point B.

3.1.3 Index system and principle of the mine roof disaster forecasting

3.1.3.1 Critical forecasting index system of the mine roof disasters

Through studying the variation law and the critical characters of the early warning indexes such as the rebound compression, bed separation displacement, loading of the supporting body and geographical information, the comprehensive weighting method is adopted to determine the weight of the early warning indexes. Then, the critical value of the corresponding early warning indexes is calculated. Finally, the critical early warning index system is formed.

Determination of the weight of the early warning indexes:

$$\begin{cases} w_n = (1 + \sum_{k=2}^{n} \prod_{i=k}^{n} r_i)^{-1} & \text{G1 method} \\ w_i = \frac{1}{n - E}(1 - e_i) & \text{Entropy weight method} \end{cases}$$

Comprehensive weighting evaluation model:

$$\begin{cases} \min H = \sum_{j=1}^{n} \alpha \sum_{o=1}^{l} (w_j - w_{oj})^2 + \sum_{j=1}^{n} \beta \sum_{s=l+1}^{q} (w_j - w_{sj})^2 \\ o.s. \sum_{j=1}^{n} w_j = 1; \; 0 \leqslant w_j \leqslant 1; \; 1 \leqslant j \leqslant n \end{cases}$$

3.1.3.2 The critical early warning method for the mine roof disasters

Analysing the space-time problem which is generated by the mine roof disaster, the space-time model of the critical warning of the mine roof disasters is constructed, as shown in Figure 3-13.

The automatic recording instrument of the multiple basement points which has the real-time distinguishing function and the roof collapsing monitoring early warning instrument are developed, as shown in Figure 3-14. This is to discover and early warn the roof collapsing potentials in which the roof displacement and the supporting body are loaded unusually.

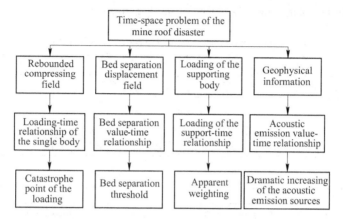

Figure 3-13   The space-time model of the critical early warning of the mine roof disasters

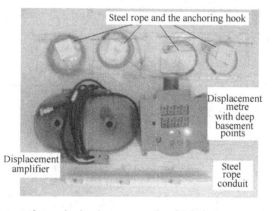

Figure 3-14   Automatic monitoring instrument of multiple basement point displacement

3.1.3.3 Early warning and forecasting expert system for the roof disasters in the mine site

The main function of the early warning system for mine site roof disasters is to collect the observable information of the early warning index system of the kinds of roofs in the mine sites. Furthermore, the index standardised processing is conducted on it and the early warning distinguishing is performed.

A The comprehensive early warning model in the roof disaster early warning system in the mine sites

On the basis of the in-situ conditions, first, the district early warning in the macroscopic area of the mine longwall face is conducted. After the classification of the district early warning is generated, the technical measure intervening is conducted on the instability type. Then, immediate early warning during the process of mining can be conducted on the types that the main roof is stable and the instable roof after the intervening is performed. The comprehensive early warning model can be established, as shown in the Figure 3-15.

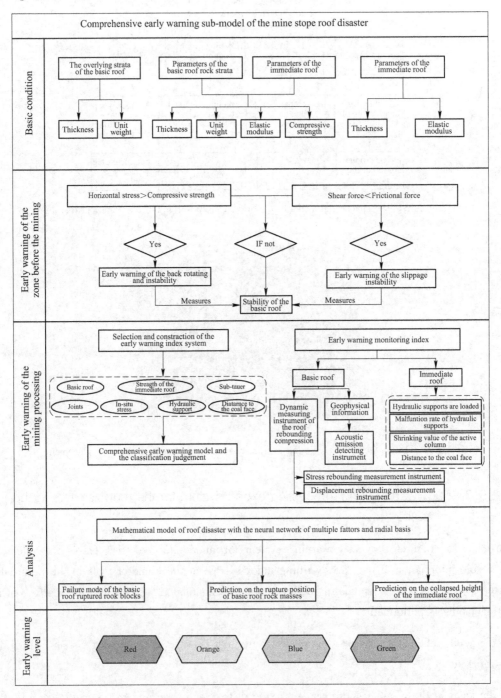

Figure 3-15 The comprehensive model of the roof disaster early warning in the mine site

B The function of the software monitoring and the early warning of the system disaster

For the mine roof disaster monitoring and early warning software, it is installed in the hydraulic support pressure transducer and the roof displacement transducer in the production field of the mine site working face. Then, in the production process, the hydraulic support pressure, the dynamic sinking value of the roof and the other early warning index data are collected. Furthermore, they are transported in time to the data base in the server which is located on the ground. The home screen of the system software is showing in the Figure 3-16.

Figure 3-16 The home screen of the monitoring and disaster early warning system

3.1.4 Demonstration engineering of the project application

The Jinhuagong Coal Mine belonging to the Datong Coal Mine Group is a complicated mine site with multiple coal seams. This mine mainly exploits the coal seams 7-3#, 7-4#, 11# and 12# (The distance between the rock layers is 8.4m, 38.8m and 19.9m). In the exploiting practices, the blocky roof of the fully mechanised working face show roof disaster controlling issues, such as thesevere roof collapsing and collapsing of the sides. The coal face roof collapsing disaster in the fully mechanised working face in Jinhuagong Coal Mine is shown in Figure 3-17.

Figure 3-17 The coal face collapsing disaster of the Jinhuagong fully mechanised working face

For the Tongxin Coal Mine, the fully mechanised working face with large mining height cuts the extremely thick coal seam with a thickness of 15m at one time. The accident of large area roof leaking and collapsing occurs, as shown in Figure 3-18. This leads to the consequence that the working face stopped production for 15 working days. This seriously influences the safety production of the mine site, leading to enormous economic loss.

Figure 3-18   The roof collapsing disaster of the fully mechanised working face with large mining height in the Tongxin Coal Mine

Through revealing the instability mechanism of the blocky roof in the working face, as shown in Figure 3-19, the periodic weighting law and characters of the large area hanging roof during the exploiting process of multiple coal seams are classified. Then, the main influencing parameters of the mine roof disasters are acquired. The comprehensive early warning index model and the corresponding early warning system are constructed.

The project forms the roof disaster early warning system, the controlling theory and technologies of the fully mechanised working face in which the malfunction detecting of the hydraulic supports, the monitoring of the supporting body and the coal face roof controlling are integrated together. Figure 3-20 shows the malfunction monitoring of hydraulic supports and thesupporting effect of fully mechanised working face is shown in Figure 3-21.

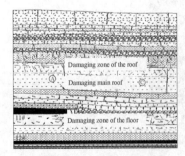  

Figure 3-19   Roof model in exploiting multiple coal seams   Figure 3-20   Malfunction monitoring of hydraulic supports   Figure 3-21   Supporting effect of fully mechanised working face

3.2 Study of the ground pressure law and the support rationalisation of the extremely large-scale fully mechanised top coal caving working face

Coal is regarded as the main energy in China. It plays a significant strategic position in the economic construction of the nation. The development of the extremely large-scale mines represents the development direction of the coal industry of the current world. In China, the coal industry is playing the advance and demonstration impact. In the area of the safety and efficient mining, the extremely large-scalemines have incomparable superiority. Meanwhile, they are subjected to the special particularity and complexity in terms of mining technology, ground pressure law and large-scale support equipment.

3.2.1 Geological production condition of the Anjialing underground coal mine 2#

For the Anjialing underground coal mine 2#, the average thickness of the coal seam 9# is 11.69m and the average dip angle of the coal seam is 2.8 degrees. The immediate roof is mainly composed of the medium coarse sandstones. Sometimes, it is composed of the fine sandstones. In the local area, there are ash black mudstones. For the main roof, it is the fine sandstone with a thickness of 8.7m. Meanwhile, it is the inferior key strata. The main key strata are the medium sandstones with a thickness of 10.2m. The immediate floor is mainly composed of muddy rocks.

The mining method of the working face B906 is fully mechanised inclined longwall top coal caving with full collapsing of the roof along the retreat direction. The mining height is 3.3m and the top coal caving height is 8.4m. The working face length along the inclined direction is 300.5m and the exploitable length along the strike direction is 1610m. The plane graph showing the layout of the roadways of the working face B906, is shown in Figure 3-22.

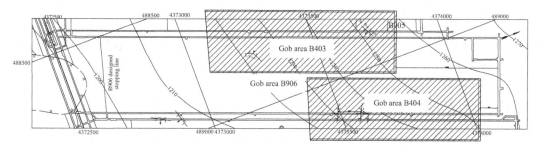

Figure 3-22　Plane graph showing the layout of the roadways of the working face B906

Through conducting the in-situ surveying and studying in the Anjialing underground mine 2#, the unusual ground pressure appearance of the fully mechanised caving working face mainly include:

(1) In the coal mine, multiple fully mechanised working faces have shown the phenomenon of the roof collapsing and the coal rib spalling. For the coal mass and rock mass in the specific fully mechanised working face, severe accidents of roof collapsing and rib spalling occurred. In the exploiting process of the working face B907, large roof collapsing and rib spalling occurred. The collapsed height is more than 8m and the depth of the rib spalling is 1~2m. For the coal and rock

mass in the fully mechanised working face B906, multiple small-range severe roof collapsing and rib spalling phenomena have occurred.

(2) Severe damaging of hydraulic supports occurred in some parts of the fully mechanised working faces. In the working face B405 and the working face B902, the phenomenon of the four-linked bar crushing failure in the middle hydraulic supports occurred. In the working face B405, 8 hydraulic supports were crushed to failure and in the working face B902, 4 hydraulic supports were crushed to failure. When the working face B906 passed through the roadway, the phenomenon that more than ten hydraulic supports were crushed tightly occurred. For the vertical columns of hydraulic supports, the phenomenon that the sealed ring heaved and the liquid was leaked occurred. As for the lifting jack, the malfunction of compressing to bending, crushing to rupture and leakage of liquid occurred.

(3) In the fully mechanised caving working face, parts of hydraulic supports showed abnormal condition. In the fully mechanised working face B906, the geometric state of the hydraulic supports is abnormal (the phenomenon such as the squeezing of hydraulic supports, raising and yielding of the top beam, and the top beam stairs); The suffered load of the front and rear vertical columns is not balanced. Also, the phenomenon that the supporting force of hydraulic supports cannot be fully developed.

Although the chock-shield-type supports with four columns that are being used in the fully mechanised top coal caving working face in the Anjialing underground coal mine 2# basically realises the safe and high efficient exploiting of the mine, the situation that the ground pressure appearance is abnormal still exists. To make the supporting equipment more economic and rational, it is necessary to deeply study the ground pressure law of the fully mechanised caving working face. According to the relationship between the hydraulic supports and surrounding rock masses, the study to modify and improve the hydraulic supports is conducted.

3.2.2 Measuring of the floor pressure ratio and the adaptability of the hydraulic support basement

3.2.2.1 Measuring principle and equipment of the floor pressure ratio

The pressure that the hydraulic support basement applies on the floor with a unit area is called as the load intensity of the floor, namely the floor pressure ratio. Figure 3-23 shows the schematic diagram showing the portable floor pressure ratio instrument (DZD-IIA).

3.2.2.2 In-situ test of the floor pressure ratio in the fully mechanised caving section

In the fully mechanised caving working face B906, there are totally 8 observation stations and 14 observation points. The in-situ test results of the floor pressure ratio are shown below:

(1) In the fully mechanised caving section of the coal seam 9#, the floor pressure character curve is mainly composed of the plastic-brittle character and the plastic deformation.

(2) The floor of the coal seam 9# has relatively higher critical compressive strength and

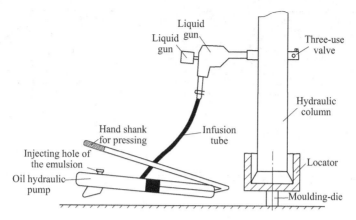

Figure 3-23 The schematic diagram showing the portable floor pressure instrument (DZD-IIA)

stiffness. The scatter of the observed results is all relatively smaller.

(3) To avoid that the hydraulic basement enters the floor, the required pressure ratio that the basement applies on the coal (rock) floor q_c and the required stiffness coefficient K_c can be calculated with the following equations:

$$q_c = cq_m = 0.75 \times 26.6 = 19.6 \text{MPa} \tag{3-8}$$
$$K_c = cK_m = 0.75 \times 4.5 = 3.4 \text{MPa/mm} \tag{3-9}$$

In the equations, c is the reserve factor and $c = 0.75$.

3.2.2.3 Evaluation on the hydraulic support basement adaptability based on the floor pressure ratio

In the design and selection of the hydraulic supports, the pressure ratio of the basement on the floor front is a significant parameter. When the basement front pressure ratio q_1 is larger than the compressive strength of the floor, the basement front will enter the floor, which will influence the movement of the hydraulic supports. However, the rear-end pressure ratio q_2 will generally not influence the movement of the hydraulic supports. In the technical characters of hydraulic supports, the front pressure ratio q_1 is more important than the average pressure ratio. The calculating parameters of the hydraulic support basement pressure and the distribution form are shown in Figure 3-24.

Considering the influence of the overlying rock strata impact load, the complexity of the hydrogeological condition in the working face and the influence of the complicated geological structure in the mining process of the fully mechanised working face, such as the faults and folds, the adaptability of the hydraulic supports on the floor is also changing. This usually makes that the rear-end pressure ratio of the hydraulic support basement is 0. The position of the hydraulic support basement resultant force acting point moves forward. At this time, $2L/3 < L_1 < L$, as shown in the Figure 3-25.

$$\sum Y = 0, q_1 = \frac{2Q}{B(L_d - m)} \tag{3-10}$$

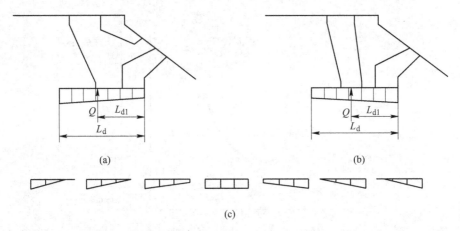

Figure 3-24 Calculating parameters and distribution form of the hydraulic support basement pressure ratio
(a) Shield top coal caving hydraulic support with two columns to support the roof;
(b) Chock-shield-type top coal caving hydraulic support with four columns; (c) Distribution form

$$\sum M_A = 0, q_1 = \frac{6QL_{d1}}{B(L_d - m)(2L_d + m)} \quad (3-11)$$

For chock-shield-type supports with four columns:

If $m = \dfrac{L_d}{4}$, $L_{d1} = \dfrac{3L_d}{4}$. Meantime, $q_1 = \dfrac{8Q}{3BL_d}$.

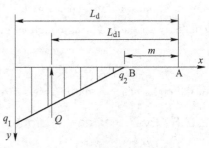

Figure 3-25 Analysis diagram of the suffered load of the hydraulic support basement

In the fully mechanised caving section of the coal seam 9# in the Anjialing underground coal mine 2#, for the sandy mudstone floor, the allowable pressure ratio is up to 19.6MPa. The allowable stiffness coefficient of the rock basement is 3.4MPa/mm. Therefore, the floor belongs to type IV in the floor classification draft, namely the medium strong type.

The fully mechanised caving working face B906 currently uses the normal four-linked bar electric controlled hydraulic support with low position to cave the top coal (ZF10000/23/37D). The width of the basement is 1.5m and the length of the basement is 3.115m. The resultant force that the hydraulic support basement suffers is calculated according to the rated working resistance 10000kN. Considering the influence of the complicated working condition, the front pressure ratio of the hydraulic support basement can be calculated as:

$$q_1 = \frac{2Q}{B(L_d - m)} = \frac{8Q}{3BL_d} = \frac{8 \times 10000}{3 \times 1.5 \times 3.115} = 5.71\text{MPa} \quad (3-12)$$

$$q = k_1 k_2 k_3 q_1 = 1.4 \times 1.3 \times 1.35 \times 5.71 = 14.1\text{MPa} < 19.6\text{MPa} \quad (3-13)$$

Where, k_1 is dynamic loading coefficient of the weighing pressure, 1.4 used; k_2 is modified coefficient of the impact dynamic load, 1.3 used; k_3 is complexity coefficient of the geological production condition, 1.35 used.

Due to the fact that $q < q_c$, it indicates that the basement of the chock-shield-type top coal caving hydraulic support with four columns that the fully mechanised caving working face B906 is used, can meet the safety production requirement. If the shield top coal caving hydraulic supports with two columns to support the roof is used, the designed value of the front pressure ratio should be controlled lower than 19.6MPa.

3.2.3 Relationship between the type of the fully mechanised top coal caving hydraulic support and the supporting performance and the rationality of the hydraulic support type

3.2.3.1 Mechanical analysis of the hydraulic support horizontal supporting force and the improvement direction

The horizontal force that the shield hydraulic support with two columns to support the roof is determined by the horizontal component of the vertical column supporting load, the maximum frictional force between the top coal and the top beam, and the maximum frictional force between the basement and the floor. When the maximum frictional force between the top beam and the top coal, the maximum frictional force between the basement and the floor are larger than the horizontal component of the vertical column supporting force, the top beam will not slip along the forward direction. When one of the maximum frictional force between the top coal and the top beam, and the maximum frictional force between the basement and the floor is smaller than the horizontal component of the vertical column, the horizontal force that the hydraulic support applies on the roof equals the smaller one of those two frictional forces.

Relevant research indicates that only 22% of the initial supporting horizontal component of the hydraulic support vertical column acts on the top coal (If the hydraulic support reaches the nominal working resistance, then only 45% acts on the top roof). This is resulted by the fact that the frictional coefficient between the top beam and the top coal is relatively smaller. If the Anjialing underground coal mine 2# uses the shield top coal caving hydraulic support with two columns to support the roof, its active horizontal supporting force can be calculated as:

$$T = k_r \eta P_1 \cos\alpha = 0.22 \times 0.75 \times 5950 \times \cos 85° = 85.6 \text{kN} \quad (3-14)$$

Overall, the horizontal force of the hydraulic supports is determined by the vertical component of the vertical columns, the horizontal component of the vertical columns, the frictional coefficient between the top coal and the roof, and the frictional coefficient between the basement and the floor. Therefore, improving the roughness of the top beam surface and the dip angle of the column leg can use the horizontal force to the maximum extent, to control the failure of the coal mass and rock mass at the coal face. Therefore, when designing the hydraulic supports, emphasis should be given.

3.2.3.2 Comparison on the bearing character of the chock-shield-type supports with four columns and the shield hydraulic support with two columns to support the roof

In the comparison between the shield hydraulic supports with two columns to support the roof and the chock-shield-type supports with four columns, there are two core pre-conditions: (1) For those two hydraulic support types, the overall matching dimension of three machines is identical; (2) The working resistance of those two types of hydraulic supports are equal, namely 10000kN.

From the Figure 3-26 and Figure 3-27, it can be seen that:

(1) Under the condition that the working resistance are all lower than 10000kN, the effective working area of the shield top coal caving hydraulic support with two columns to support the roof is mainly distributed around the vertical column. The width is 0.47m. As for the chock-shield-type top coal caving hydraulic supports with four columns, the effective working area is located between two columns. The width is 1.457m. Therefore, the adaptability that the chock-shield-type supports with four columns have on the variation of the external resultant load acting point is stronger.

(2) For the shield top coal caving hydraulic supports with two columns, the critical bearing position is located at 1.09m of the top beam end. However, for the chock-shield-type top coal caving hydraulic supports with four columns, the critical bearing position is located at 1.453m to the top beam end.

(3) In the comparison between the shield hydraulic supports with two columns to support the roof and the chock-shield-type supports with four columns, there are two core pre-conditions: 1) The length of the top beam for those two types of hydraulic supports is identical; 2) The working resistance of those two types of hydraulic supports are equal, namely 10000kN.

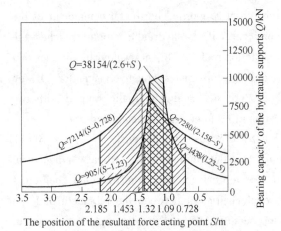

Figure 3-26 Comparison diagram showing the force balance area of the chock-cover-type hydraulic supports with four columns and the shield hydraulic supports with two columns to support the roof

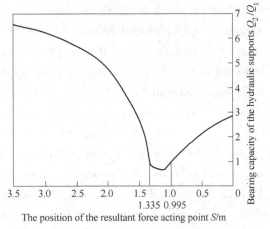

Figure 3-27 The relationship between the bearing capacity ratio of hydraulic supports (Q_2/Q_1) and the position of the resultant force acting point
Q_1—Bearing capacity of the shield hydraulic support with two columns to support the roof;
Q_2—Bearing capacity of chock-shield-type supports with four columns

3.2.3.3 Evaluation on the rationality of the fully mechanised caving hydraulic support type based on the hydraulic support performance

The advantages and disadvantages of the performance of the shield top coal caving hydraulic supports with two columns to support he roof:

Advantages:

(1) The shield hydraulic supports with two columns to support the roof can provide effective horizontal supporting force. It improves the stress distribution in the coal and rock masses in the coal face. Furthermore, it decreases the tensile stress failure zone, which is beneficial for controlling the coal and rock masses at the coal face.

(2) The phenomenon that the suffered load of the front vertical columns and the rear vertical columns are not balanced will not occur. The supporting force of the hydraulic support vertical columns can be fully developed.

(3) There are two vertical columns and the hydraulic support structure is simple. The reliability is high.

(4) The economic benefits are favourable. The weight of the hydraulic supports is light. The consumed quantity of the steel is small. The transport operation is convenient.

(5) The operation of the electric-hydraulic controlling system is more convenient. The advancing velocity is relatively faster, which is beneficial for high efficient mining.

Disadvantages:

(1) The effective working range is small. The adjusting range of the balance lifting jack is limited. The adaptability of the hydraulic supports on the variation of the roof resultant load acting point is relatively weak. When the hydraulic supports are being designed, the chock-shield-type lifting jack can be considered to overcome this shortcoming.

(2) It is not applicable for the roof condition where the main roof weighting is violent.

The fully mechanised caving working face B906 of the Anjialing underground coal mine 2# has shown the phenomenon of severe roof collapsing and rib spalling in the local area. However, the main roof weighting is not violent. To actively and reliably control the collapsing and leaking of the coal and rock mass at the coal face, and make the supporting equipment at the working face more economic and efficient, the shield top coal caving hydraulic supports with two columns to support the roof are used.

3.2.4 Design of main parameters of the fully mechanised caving shield hydraulic supports with two columns to support the roof

3.2.4.1 Calculation and determination of the reasonable working resistance of hydraulic supports

If the shield hydraulic supports with two columns to support the roof is used, the reasonable working resistance can be determined based on the following methods.

(1) Determining the working resistance of hydraulic supports through the in-situ observation:
$$P = 7803 \text{kN}$$
(2) Determining the working resistance of hydraulic supports based on the statistic theory:
$$P = 7889 \sim 8386 \text{kN}$$
(3) Determining the working resistance of hydraulic supports based on the roofclassification:
$$P = 7690 \sim 8331 \text{kN}$$
(4) Determining the working resistance of hydraulic supports based on the empirical equations:
$$P = 7417 \sim 8035 \text{kN}$$
(5) Determining the working resistance of hydraulic supports according to the commonly used equations for the top coal caving hydraulic supports:
$$P = 8083 \text{kN}$$

Based on the above kinds of calculating methods, it is determined that for the shield top coal caving hydraulic supports with two columns to support the roof, when the width is 1.5m, the working resistance is 8500kN and the initial supporting force is 5950kN. When the width is 1.75m, the working resistance is 10000kN and the initial supporting force is 7000kN.

3.2.4.2 Determination of the main structural parameters of the hydraulic supports

(1) Determination of the hydraulic support height.
$$H_m \geqslant h_m + m_1 \geqslant 3.95 \text{m} \tag{3-15}$$
$$H_n \leqslant h_n - m_2 - a - \sigma_a \leqslant 2.05 \text{m} \tag{3-16}$$

(2) Determination of the hydraulic support spacing.

For the fully mechanised caving working face B906, the length of each section in the middle groove of the scraper conveyor is 1.5m. The connecting block position of the lifting jack is in the centre of the middle groove. The hydraulic support spacing is 1.5m.

(3) Determination of the top beam parameters.

The top beam length = The mating dimension + the middle centre spacing of the rear conveyor + the width of the front scraper conveyor/2 + the width of the rear scrape conveyor/2 - ($G\cos\beta$ + 300 + e) + the distance from the junction point of the shield beam and the top bean to the rear end of the top beam. After calculation, it is 4910mm.

(4) Calculation of the roof controlling distance.

After calculation, the maximum roof controlling distance is 6.26m and the minimum roof controlling distance is 5.46m.

(5) Determination of the hydraulic supporting area.
$$F_c = b_c(L_g + \Delta) = 9.39 \text{m}^2 \tag{3-17}$$

(6) Determination of the technical parameters of the vertical columns.

The top nest position of the vertical column is calculated as shown in Figure 3-28.
$$S_1 = \frac{1}{Q_1}(L_1 P_1 \cos\alpha_1 + L_2 P_2 \cos\alpha_2) = 1.54 \text{m}, S_2 = 1.44 \text{m} \tag{3-18}$$

Considering that the suffered load of the vertical column of the hydraulic supports with four

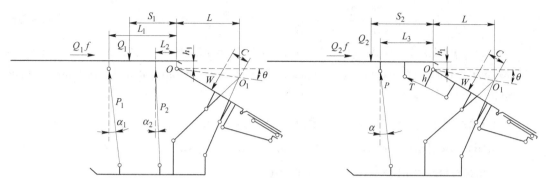

Figure 3-28 The calculation diagram showing the top beam nest of the vertical column

columns at the top coal caving working face is not balanced, the resultant force acting point of the hydraulic support with two columns should move forward. Combining with the condition that the roof weighting is apparent, according to Figure 3-29, it is determined that the distance from the centre of the top column nest of the vertical column to the junction point of the top beam and the shield beam is:

$$L_3 = 1.35 \text{m}$$

The distance from the bottom column nest centre of the vertical column to the junction point between the basement and the rear linked bar is:

$$L_4 = L_3 + G\cos\beta - A\cos\alpha_3 - H_3\tan\alpha = 1.65\text{m} \tag{3-19}$$

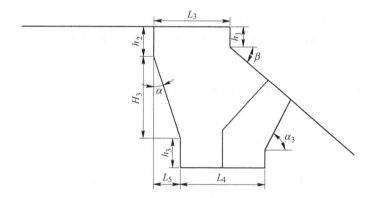

Figure 3-29 Calculation diagram showing the position of the bottom column nest

3.2.4.3 Design of the basement parameters based on the floor pressure ratio

According to the requirement of the in-situ production condition and the matching equipment dimension, it is determined that the hydraulic support basement parameters are that the basement width $B = 1.5$m and the basement length $D = 3.2$m. Under the condition that the working resistance is 8500kN, the floor pressure ratio of the basement for the shield hydraulic support with two columns to support the roof is:

$$q_1 = \frac{2Q}{B(L_d - m)} = \frac{3Q}{BL_d} = \frac{3 \times 8500}{1.5 \times 3.2} = 5.31 \text{MPa} \qquad (3\text{-}20)$$

$$q = k_1 k_2 k_3 q_1 = 1.4 \times 1.3 \times 1.35 \times 5.31 = 13.1 \text{MPa} < 19.6 \text{MPa} \qquad (3\text{-}21)$$

Through the numerical calculation examination, it can be acquired that the dimension of the hydraulic support basement can fulfil the requirement of the floor pressure ratio.

3.3 Controlling of the hydraulic support-surrounding rock mass in the fully mechanised top coal caving working face with large dip angle under the complicated condition

In China, the reserves of the coal seam with the dip angle larger than 35 degrees accounts for one fifths of the total reserves. However, its production is smaller than one tens of the whole coal production of China. In the middle and east area, the coal seams with small dip angle are going to be exhausted. Exploiting the coal seams with large dip angle becomes the compulsory condition for the sustainable development.

In the west area, the main exploiting coal seams of a number of mine sitesare coal seams with large dip angle. The problems of the roof collapsing at the coal face and the downward slippage of the hydraulic supports at the fully mechanised working face with large dip angle, not only threaten the safety of coal operators, but also lead to the large consumption of the labour, wealth and stuff, and the large engineering maintaining quantity. This restrains the safety and high efficient production of mines. Therefore, the efficient controlling of the hydraulic supports and surrounding rock masses of the fully mechanised working face with large dip angle under the complicated condition, has already become the scientific challenge that needs to be solved urgently in the safe and high efficient production of coal mines.

3.3.1 Engineering profile of the fully mechanised caving working face with large dip angle under the complicated condition

3.3.1.1 In-situ geological profile

The fully mechanised caving working face 2313 is located in the south of the west wing in the mining area 230. The average thickness of the coal seam is 6.50m and the maximum dip angle of the working face is up to 51 degrees. The average dip angle is 36 degrees. The immediate roof is the sandy mudstone, which has the thin layer profile or the medium thick layer profile. In the local area, there are horizontal beddings. The north area is mainly composed of siltstones and fine sandstones, which have the grey white colour. It is argillaceous and bonded, which has high strength. The main roof is composed of medium sandstones, which has the thick layer profile. In the local area, it has the thin layer profile. And it is mainly composed of quartz. The unregular fine sandstone combined body is included and the strength is relatively higher. It also has fractures. The immediate floor is mudstones. In the local area, there are fine sandstones and the strength is relatively smaller. It also has the uneven fracture. According to the roadway tunnelling and the revealing situation in the working face mining period in the section 2313, there are 4 faults whose vertical falling is larger than 2m. Also, it is revealed that there are 4 faults whose vertical falling is smaller than 2m.

3.3.1.2 In-situ production condition

The fully mechanised caving working face 2313 uses the fully mechanised top coal caving retreat mining method with the longwall face along the dip direction. The coal shearer with two rollers (MG150/368-WD) is used to cut coal. The mining height is 2.3m and the coal caving height is 4.2m. For the coal caving, one caving with two coal cutting is used. The coal caving with intervals is conducted between multiple cycles. The coal caving step is 1.2m.

The type of the front and rear scrape conveyoris SGZ-630/264. A reversed loader with the bridge type is used and its type is SZZ-764/132.

For the main hydraulic supports in the fully mechanised working face2313, 47 top coal caving hydraulic supports with low caving position (ZF4200/16/26) is selected. As for the top end and bottom end, 5 transition hydraulic supports (ZFG4200/16/26) are used (There are 3 hydraulic supports at the top end and 2 hydraulic supports at the bottom end).

3.3.1.3 Character and the reason of the hydraulic support collapsing accident in the roof collapsing

The accidents happened in the fully mechanised caving working face in the Ge'ting Coal Mine are mainly composed of roof collapsing and hydraulic support collapsing. They are mainly reflected in the following aspects:

(1) The working face with large dip angle 2313 and other fully mechanised top coal caving working face have shown the roof collapsing accidents with the collapsed height up to 2.5m. In the local area, the collapsed coal and rock masses at the coal face gushed to the outside of the scrape conveyor and the inside of the hydraulic supports. This leads to the low supporting efficiency of hydraulic supports. This further leads to the cycle of the further degradation of the roof collapsing.

(2) For the fully mechanised working face with large dip angle 2313, due to the fact that the coal seam dip angle is large, large scale inclination and collapsing accidents of hydraulic support occurred. The number of inclined and collapsed hydraulic supports account for 50% of the total number of hydraulic supports in the working face. This leads to the termination of the working face and it is forced to develop a new working face in the outside area.

The reasons leading to theroof collapsing and hydraulic support collapsing accident in the fully mechanised working face with large dip angle mainly include the following six aspects:

(1) In the fully mechanised caving working face, the quantity of the fault structure is large. This is easy to lead to the result that the top coal is fractured and the roof contacting of the hydraulic supports is uneven.

(2) The coal mass strength is low. The joints and fractures are developed. This is easy to

generate the phenomenon that the coal and rock masses are leaked and collapsed, and the coal rib is spalled.

(3) For the hydraulic supports at the fully mechanised working face with large dip angle, the suffered load status is complicated. It is difficult to be controlled.

(4) The self-weight of the original hydraulic supports is large. When the position and status are not right, it is relatively difficult to be adjusted.

(5) The old hydraulic supports have multiple malfunction in the hydraulic system, leading to the consequence that the reliability of the hydraulic support roof controlling is poor.

(6) The exploiting and the operating management of the fully mechanised top coal caving working face with large dip angle are difficult. There are several production and safety hidden dangers.

3.3.2 The hydraulic support collapsing accident with large dip angle and its controlling parameters

3.3.2.1 Instability reason of the hydraulic supports with large dip angle

The instability reason of the hydraulic supports with large dip angle mainly include five different aspects, such as the geological condition of the fully mechanised caving working face, the character of hydraulic supports, the characters of the surrounding rock masses in the fully mechanised caving working face, the coal mining method of the fully mechanised caving working face and the staff management factor. The specific instability reasons are shown in Figure 3-30.

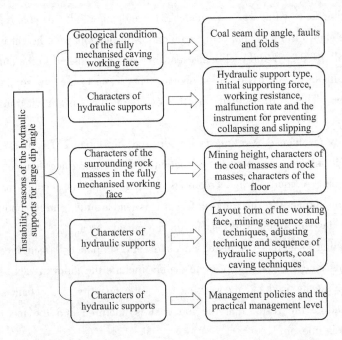

Figure 3-30 Instability reasons of the hydraulic supports with large dip angle

3.3.2.2 Mechanical analysis on the stability of the hydraulic supports with large dip angle

A Stability analysis of the hydraulic supports under the roof (coal) collapsing condition

Under the roof (coal) collapsing condition, the suffered force of the fully mechanised caving hydraulic supports is mainly composed of two parts, namely the self-weight of the hydraulic supports and the acting force that the floor applies on the bottom pedestal of the hydraulic supports. As shown in Figure 3-31, G is the weight of the hydraulic support, kN; W_1 is the normal pressure that the floor applies on the bottom pedestal of the hydraulic support, kN; α is the dip angle of the coal seam, (°); μ_1 is the frictional coefficient between the bottom pedestal of the hydraulic support and the floor; H is the practical supporting height of the hydraulic support, m; H_g is the height of the gravity centre for the hydraulic support, m; B is the width of the bottom pedestal for the hydraulic support, m.

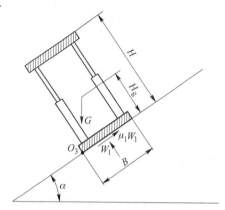

Figure 3-31 The mechanical model of the hydraulic support falling under the roof (coal) collapsing condition

The critical coal seam dip angle when the hydraulic support shows rotating collapsing under the effect of the self-weight is:

$$\alpha_{m1} = \arctan \frac{B}{2H_g} \qquad (3\text{-}22)$$

The critical coal seam dip angle when the hydraulic support shows downward slippage under the effect of the self-weight is:

$$\alpha_{m2} = \arctan\mu_1 \qquad (3\text{-}23)$$

B Stability analysis of the hydraulic supports under the condition that the hydraulic support is moved upon contacting the roof

Figure 3-32 and Figure 3-33 show the cross-section diagram along the strike direction and the cross-section diagram along the dip direction of the fully mechanised hydraulic support suffered load during the hydraulic support moving process.

The minimu value of the vertical column supporting force to guarantee the forward moving of the hydraulic support with contacting the roof is:

$$F_{m1} = \frac{Q_m(l_4 - f_2 l_5) + (G_2 l_{g2} - G_1 l_{g1})\cos\alpha}{L_4 - L_1\cos\theta_1 + \lambda L_3 - \lambda L_2\cos\theta_2} + \frac{\gamma Bhm\cos\alpha(l_3 + f_3 l_6)}{L_4 - L_1\cos\theta_1 + \lambda L_3 - \lambda L_2\cos\theta_2} \qquad (3\text{-}24)$$

The minimum value of the vertical column supporting force to prevent the downward slippage of hydraulic supports is:

$$F_{m2} = \frac{G\sin\alpha(1-\mu_1)(l_4 - f_2 l_5)}{2(\mu_1 + \mu_2)(L_4 - L_1\cos\theta_1 + \lambda L_3 - \lambda L_2\cos\theta_2)} +$$
$$\frac{(G_2 l_{g2} - G_1 l_{g1})\cos\alpha + \gamma Bhm\cos\alpha(l_3 + f_3 l_6)}{2(L_4 - L_1\cos\theta_1 + \lambda L_3 - \lambda L_2\cos\theta_2)} \quad (3-25)$$

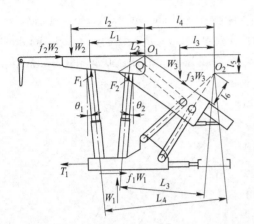

Figure 3-32　The cross-section diagram showing the fully mechanised hydraulic support suffered load during the hydraulic support moving process

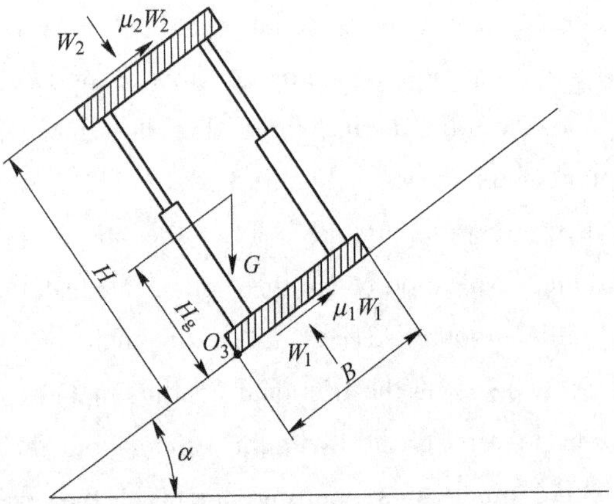

Figure 3-33　The cross-section diagram showing the fully mechanised hydraulic support suffered load along the dip direction during the hydraulic support moving process

The minimum vertical column supporting force to prevent the hydraulic support collapsing is:

$$F_{m3} = \frac{G(2H_g\sin\alpha - B\cos\alpha)(l_4 - f_2 l_5)}{2(B + 2H\mu_2)(L_4 - L_1\cos\theta_1 + \lambda L_3 - \lambda L_2\cos\theta_2)} +$$
$$\frac{(G_2 l_{g2} - G_1 l_{g1})\cos\alpha + \gamma Bhm\cos\alpha(l_3 + f_3 l_6)}{2(L_4 - L_1\cos\theta_1 + \lambda L_3 - \lambda L_2\cos\theta_2)} \quad (3-26)$$

The critical coal seam dip angle α_m to guarantee that the hydraulic support does not collapse and slip when the anti-collapsing and anti-slipping measures have not been performed is:

$$\alpha_m = \min\{\alpha_{m1}, \alpha_{m2}\} \quad (3-27)$$

The minimum vertical column supporting force F_m to guarantee that the hydraulic supports do not collapse and slip is:

$$F_m = \max\{F_{m1}, F_{m2}, F_{m3}\} \quad (3-28)$$

3.3.2.3　Selection and optimisation of the hydraulic support with large dip angle

A　Comparison and selection of two hydraulic supports

The performance comparison of the hydraulic supports ZF5600/16.5/26 and ZF4200/16/26 is shown in Table 3-2.

Through the above comparison analysis and the practical situation of the Ge'ting Coal Mine, it can be acquired that the fully mechanised caving working face 2313 with large dip angle is not appropriate to use the heavy hydraulic support (ZF5600/16.5/26). In fact, the hydraulic support ZF4200/16/26 should be selected.

Table 3-2 Performance of hydraulic supports ZF5600/16.5/26 and ZF4200/16/26

Comparison	Type of the hydraulic support	ZF5600/16.5/26	ZF4200/16/26
1	Weight	19.8 tons (heavy)	12.5 tons (light)
2	Anti-slippage tank and hydraulic support advancing step	Not coordinated	Coordinated
3	Shield effect of the rear beam on the rear scrape conveyor	Cannot fully cover when the angle is more than 40 degrees	The shield beam and the rear bean can relatively cover better
4	The rear space behind the hydraulic support	Small	Large
5	The rear anti-slippage and the position of the bottom tank	Not coordinated and being easy to be damaged	Coordinated
6	Stroke of the flank plate	Small (43mm)	Large (55mm)
7	Anti-collapsing and anti-slippage equipment	Yes	Should be modified and then installed
8	Supporting force when moves the hydraulic support with contacting the roof	334kN (Large)	223kN (Small)
9	Selection of the hydraulic support type	Working face with small dip angle	Working face with large dip angle

B Controlling parameter calculation of the hydraulic support with large dip angle

The critical coal seam dip angle (α_m) to guarantee that the hydraulic support ZF4200/16/26 in the fully mechanised caving working face 2313 does not collapse and slip under the condition that the roof is collapsed is:

$$\alpha_m = \min\{\alpha_{m1}, \alpha_{m2}\} = 15.6° \quad (3\text{-}29)$$

The minimum vertical column supporting force to guarantee that the hydraulic supports at the working face 2313 do not collapse and slip when the dip angle is 51 degrees is F_m. Then, the minimum vertical column supporting force to prevent the hydraulic support ZF4200/16/26 from collapsing and slipping is:

$$F_{m4200} = \max\{F_{m1}, F_{m2}, F_{m3}\} \approx 223\text{kN} \quad (3\text{-}30)$$

The minimum vertical column supporting force to prevent the hydraulic support ZF5600/16.5/26 from collapsing and slipping is:

$$F_{m5600} = \max\{F_{m1}, F_{m2}, F_{m3}\} \approx 334\text{kN} \quad (3\text{-}31)$$

From the calculating results, it can be known that the hydraulic support ZF5600/16.5/26 needs more supporting force when compared with the hydraulic support ZF4200/16/26. The adjusting is more difficult.

Anti-slippage design of the top coal caving hydraulic support with low caving position (ZF4200/

16/26) is shown in Figure 3-34 and Figure 3-35. To be more specific:

(1) The tubes are installed in the internal square platform in the rear of the hydraulic support basement. In the tubes, trunnions that can rotate along the left and right side are installed. For every two hydraulic supports, a lifting jack with $\phi 100$mm is used to be fixed on the trunnions that are added on the hydraulic supports. This is to realise the adjusting of the hydraulic support angle.

(2) For the middle hydraulic support in the working face, the trunnions are installed in the basement. Every 3~5 hydraulic supports, a set of anti-slippage tank that is used to pull the scrape conveyor is installed (The diameter is 140mm and the stroke is 700mm). The anti-slippage tank is used to adjust the scrape conveyor, preventing the downward slippage of the scrape conveyor which will lead to the slippage of hydraulic supports.

(3) Every 5~10 hydraulic supports, a compressing tank is installed (the diameter is 140mm and the stroke is 700mm). The top of the tank is installed on the top beam of the hydraulic supports. The bottom extends, which can be compressed on the pushing connecting rod. This makes it contact the floor tightly and it will not raise. This prevents the lateral rolling of the scrape conveyor that is connected with the pushing rod.

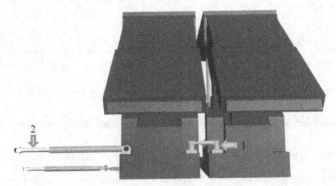

Figure 3-34 The schematic diagram showing the adjusting tank in the rear of the hydraulic supports and the rear scrape conveyor pushing rod
1—Oil cylinder in the bottom; 2—Pulling the oil cylinder of the rear scrape conveyor

Figure 3-35 The schematic diagram showing the pushing rod tank of hydraulic supports and the tank which pulls the front scrape conveyor
1—Oil cylinder of the pressure lever; 2—Pulling the oil cylinder of the front scrape conveyor

3.3.3 Leakage malfunction detecting technology of the hydraulic supports

The malfunction analysis of the hydraulic system is shown as following:

(1) In the reasons leading to two types of malfunctions of hydraulic support vertical column and lifting jack, a small part is the malfunction of the hydraulic system such as the pump and the pipes, and the mechanical malfunction of the lifting jack. A large part is leaded by the malfunction of the hydraulic valve.

(2) The main reason leading to the malfunction of the hydraulic valve is the abrasion of the set of valves, obsolescence and the loose sealing of the sealing element. This leads to the leakage failure of the sealing ring and the interlock element of the valve.

The influence of the hydraulic support leakage malfunction on the vertical column supporting force is mainly related to two factors, namely the leakage extent and the leakage time. Figure 3-36 shows the relationship between the hydraulic support supporting force and the time with different malfunction extent.

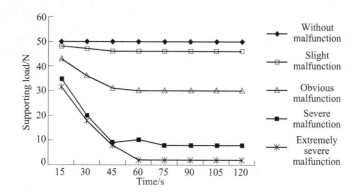

Figure 3-36 The relationship between the supporting force of the hydraulic support and the time with different malfunction extent

The leakage of hydraulic supports is divided into the "external" leakage and the "internal" leakage. The "external" leakage indicates the surface of the hydraulic supports. The sealing of the hydraulic system that can be seen is not favourable. Or, the emulsion leaks to the outside area, which is resulted by the breakage. The "internal" leakage indicates the internal area of the hydraulic system of the hydraulic supports. The components that cannot be seen such as the valve, the vertical column and the lifting jack have the internal leakage and the liquid string. The laboratory simulation is conducted to simulate the leakage of the hydraulic support hydraulic pressure. It is acquired that the range of the leaked frequency band is 3~19kHz. The ultrasonic transducer and the acceleration sensor are used respectively to detect the "external leakage" and the "internal leakage". The generated frequency signals when the hydraulic supports have malfunctions are shown in Figure 3-37.

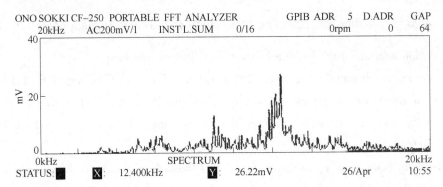

Figure 3-37 The signal diagram of the generated frequency when there is malfunction

As shown in the Figure 3-38, it is the newly developed hydraulic pressure leakage malfunction detecting instrument YHX. In the fully mechanised caving working face 2313, the hydraulic support failure malfunction detecting practice results show that:

(1) The YHX type detecting instrument has the malfunction positioning function. This shortens the time to find the leakage function and improves the hydraulic support management level in the field.

(2) The YHX type leakage detecting instrument can individually detect the leakage malfunction of the hydraulic pressure system. It can improve the definite diagnosis rate.

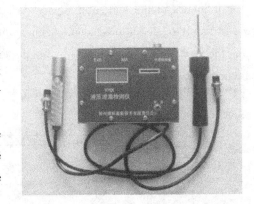

Figure 3-38 The hydraulic pressure leakage malfunction detecting instrument with the YHX type

(3) The YHX type leakage detecting instrument is appropriate for the severe production working environment. Furthermore, it has the characters that it is convenient for taking. Furthermore, the cost is low and the detecting reliability is high.

(4) In the fully mechanised caving working face 2313, the positive cycle that in-situ detecting → in-situ rectification → analysis on the ground → releasing and feedback of the special report → processing by the fully mechanised caving team is formed.

3.3.4 Safety mining controlling practices in the fully mechanised caving working face with large dip angle in the Ge'ting Coal Mine

3.3.4.1 The ground pressure observation in the fully mechanised caving working face with large dip angle

The ground pressure observation station layout of the fully mechanised caving working face with large dip angle is shown in Figure 3-39. The observation contents, purposes and measures are shown in Table 3-3.

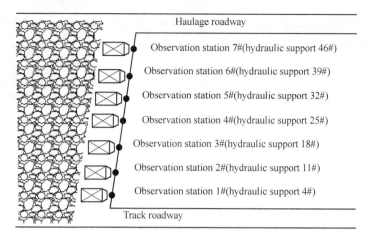

Figure 3-39 The ground pressure observation station layout diagram of the fully mechanised caving working face with large dip angle

Table 3-3 Observation contents, purposes and measures

Number	Observation contents	Observation purposes	Observation measures
1	The hydraulic pressure leakage status of the hydraulic support vertical column and plate valve	Detecting the malfunction of the hydraulic support system	The leakage detecting instrument of the hydraulic support YHX
2	The hydraulic pressure information of the hydraulic support vertical column	Distinguishing and analysing the ground pressure law of the working face	The automatic monitoring system of the hydraulic support (KJ216) pressure information
3	The hydraulic support vertical column dip angle, the pitch angle of the top beam, the pushing and transport intersection angle and the dip angle of the scrape conveyor	Distinguishing the geometric position status of the hydraulic supports	Tilt gauge and the compass
4	The mining height, the roof controlling distance, the collapsed height and range, the coal rib spalling depth and the range, the top beam stair	Distinguishing the roof stability and the hydraulic support working status	Steel tape and steel ruler

The detecting results of the leakage detecting instrument for the hydraulic support YHX is shown in Table 3-4.

Table 3-4 Statistical table of the hydraulic support leakage malfunction

Malfunction extent	The 1st detecting period			The 2nd detecting period			The 3rd detecting period		
	Malfunction position			Malfunction position			Malfunction position		
	Plate valve	Valve	Vertical column	Plate valve	Valve	Vertical column	Plate valve	Valve	Vertical column
Slight	20	5	1	22	1	2	10	2	1
Relatively severe	12	3	1	6	1	0	2	1	0
Severe	10	3	1	2	2	0	2	0	0
Extremely severe	2	1	0	1	1	0	0	0	0

The in-situ observation practices of the hydraulic support leakage malfunction show that the plate valve is the malfunction position that mostly occurred. Then, it is the one-way safety valve group. Lastly, it is the vertical column.

3.3.4.2 Determination and the controlling measures of the hydraulic support-surrounding rock mass controlling index

The hydraulic support-surrounding rock mass controlling index of the fully mechanised caving working face 2313 in the Ge'ting Coal Mine is shown in Table 3-5.

Table 3-5 The core controlling index of the hydraulic support-surrounding rock masses in the fully mechanised caving working face 2313

Controlling contents	Roof controlling distance /m	Initial supporting force /kN	Working resistance /kN	The residual resistance of the hydraulic support when it is moving with pressure/kN	The deviation value of the top beam pitch angle (γ) /(°)	The deviation value of the hydraulic support vertical column dip angle (α)/(°)	The top beam stair /m	The deviation value of the pushing and transporting intersection angle/(°)
Controlling range	$\leqslant 0.75$	$\geqslant 2200$	$\geqslant 4200$	$\geqslant 223$	$0 \leqslant \gamma \leqslant 5$	$-8 \leqslant \alpha \leqslant 2$	$\leqslant 0.1$	$\leqslant 5$

Through the comprehensive study and the in-situ practices of the roof collapsing and the controlling of the hydraulic support collapsing in the fully mechanised caving working face (2313) with large mining height, the controlling measures for the roof collapsing and the hydraulic support collapsing of the fully mechanised caving working face with large mining height is acquired. They are divided into five aspects, namely the light-weight trend of the hydraulic support type and the selection of the proper parameters, the design and installation of the anti-slippage and anti-collapsing instrument, the layout form of the equipment in the working face, the operating techniques of the equipment in the working face, and the detecting and maintaining of the hydraulic support with malfunctions. The specific information is shown in the Figure 3-40.

3.3.4.3 Analysis of the roof collapsing and hydraulic support collapsing controlling in the fully mechanised caving working face 2313

After the hydraulic support-surrounding rock mass controlling measures are conducted strictly in the fully mechanised caving working face 2313, the favourable controlling effect is acquired in the fully mechanised caving working face 2313 during the whole retreat mining period. The safety and high efficient mining of the working face with large dip angle under the complicated condition is realised. The main results are described as following:

(1) Through implementing the research results of this project in the fully mechanised caving working face 2313, the fully mechanised coal mining is initially realised in the working face with large dip angle. Specifically, the averaged dip angle of the coal seam is 36 degrees and the maximum dip angle is 51 degrees.

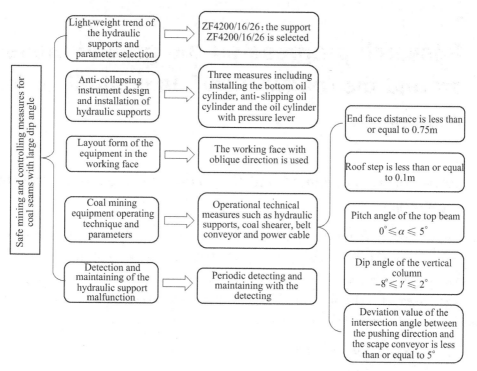

Figure 3-40 The controlling measures of the hydraulic supports-surrounding rock masses in the fully mechanised caving working face

(2) Through implementing the "the core controlling indexes of the hydraulic support-surrounding rock masses in the fully mechanised caving working face" and the "safety exploiting controlling measures in the fully mechanised caving working face with large dip angle", the maximum collapsed height of the top coal in the coal face is 0.5m. The maximum upward deviation angle of the hydraulic support vertical column along the dip direction is 8 degrees. The maximum downward angle is 2 degrees.

(3) In the service period of the fully mechanised caving working face 2313, the coal production under safety is 194048 tons. There are no accidents that influence the production, such as the collapsing and leaking of the roof, and the collapsing and downward slipping of the hydraulic supports.

4 Research progress on the ground pressure around the roadway and its controlling technology

4.1 Rock bolt reinforcement and the roadway surrounding rock mass reinforcing[9]

The coal mine roadway is a massive underground engineering. In the coal mines in China, the roadways that are need to be maintained are up to 100 thousand kilometres. The roadways that are newly tunnelled in each year is up to more than 10 thousand kilometres. The safety and high efficient reinforcement of the roadways in the coal seam rock strata is the core scientific problems in the international coal mining and geotechnical engineering.

4.1.1 Principles for rock support

(1) Philosophy for rock support.

To help the rock mass to support itself.

(2) Support categories.

1) Active support (reinforcement): Bolts, cables, shotcrete.

2) Passive support: Timber, steel sets, concrete lining, shotconcrete ribs, mesh.

(3) Principles for rock support in weak rock and in massive and structured rock.

1) Theory for weak rock——GRC (as shown in Figure 4-1).

2) Principle 1 for massive and structured rock.

① Loading condition: Weak roof is loaded by its own weight (as shown in Figure 4-2).

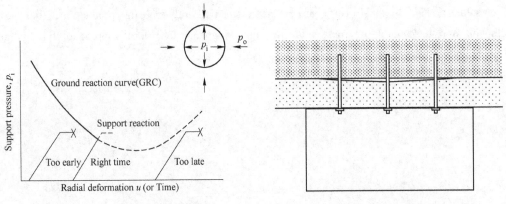

Figure 4-1 Ground reaction curve

Figure 4-2 Suspension

② Reinforcement principle: Hang up the weak roof to the competent rock above.

3) Principle 2 for massive and structured rock.
① Loading condition: The roof layers are loaded by its own weight + ground pressure (as shown in Figure 4-3).
② Reinforcement principle: Combine thin layers together, building up an interacted beam.
4) Principle 3 for massive and structured rock.
① Loading condition: Loaded by high stresses parallel with the free surface (as shown in Figure 4-4).

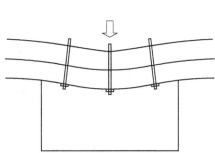

Figure 4-3 Stitching   Figure 4-4 Confinement

② Reinforcement principle: Prevent from disintegration of the failed rock.
5) Principle 4 for massive and structured rock.
① Loading condition: Relatively high ground pressure (as shown in Figure 4-5).

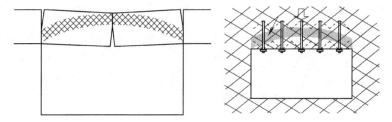

Figure 4-5 Arching

② Reinforcement principle: An arch is built up in the bolted area, which diverts the ground pressure to the abutments.
(4) Arching effect (as shown in Figure 4-6~Figure 4-8).

Figure 4-6 Field observations

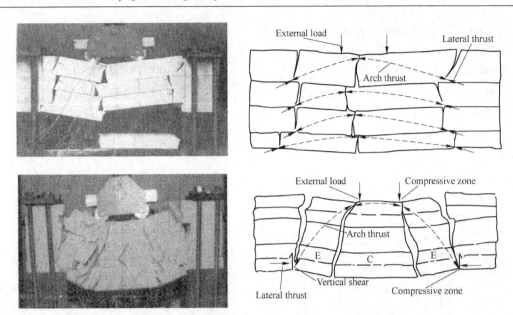

Figure 4-7 Experimental evidence (Roko et al., 1984)

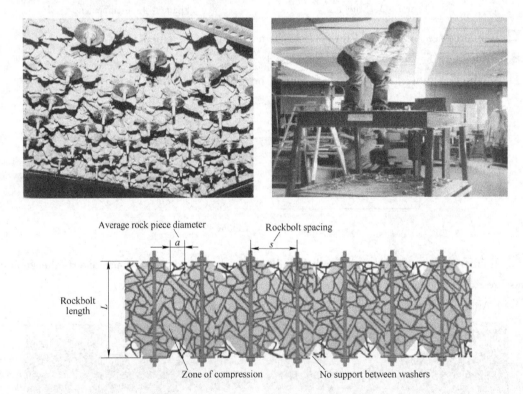

Figure 4-8 Physical model demonstrating rockbolting (Hoek, 2007)

4.1.2 Bolt-rock interaction

(1) Bolt-rock interaction——Pullout tests (as shown in Figure 4-9 and Figure 4-10).

(2) Bolt-rock interaction——In-situ measurement (as shown in Figure 4-11).

4.1 Rock bolt reinforcement and the roadway surrounding rock mass reinforcing

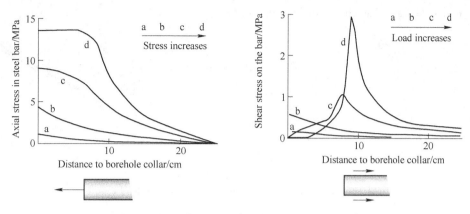

Figure 4-9 Pullout tests (Hawkes & Evans, 1951)

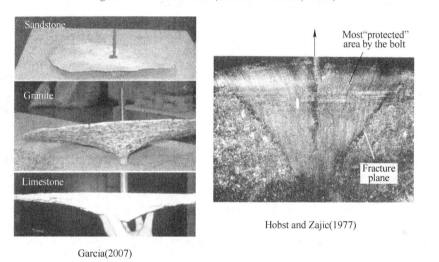

Garcia(2007)

Figure 4-10 Pullout tests

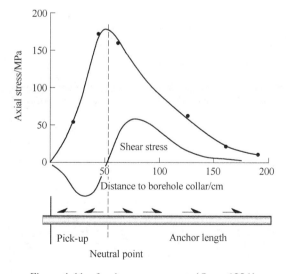

Figure 4-11 In-situ measurement (Sun, 1984)

(3) Model for bolts (as shown in Figure 4-12).

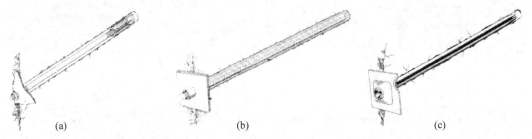

Figure 4-12 Types of bolts
(a) Mechanical bolts; (b) Fully grouted bolts; (c) Frictional bolts

1) Model for mechanical bolts (as shown in Figure 4-13).

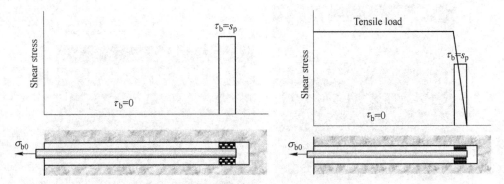

Figure 4-13 Model for mechanical bolts

2) Model for fully grouted bolts (as shown in Figure 4-14).

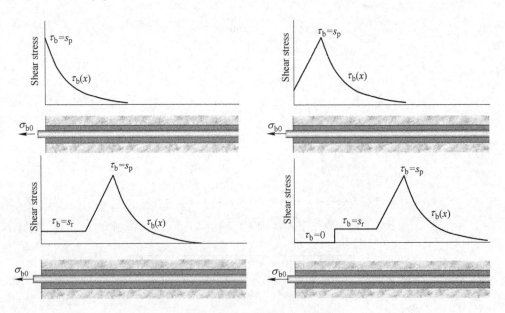

Figure 4-14 Model for fully grouted bolts

3) Model for frictional bolts (as shown in Figure 4-15).

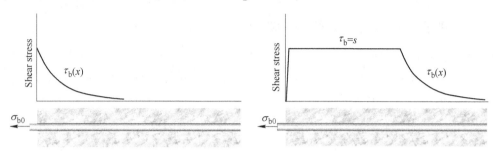

Figure 4-15 Model for frictional bolts

(4) Shear stress vs. axial stress in a bolt (as shown in Figure 4-16).

$$\tau_b = -\frac{A}{\pi d_b}\frac{d\sigma_b}{dx} \quad (4\text{-}1)$$

4.1.3 A practical problem of rebar bolts

Rock bolting adapted to rock conditions (as shown in Figure 4-17).

(1) High load-bearing capacity is required to stabilize blocks in low rock stress condition → Stiff and strong rock bolts.

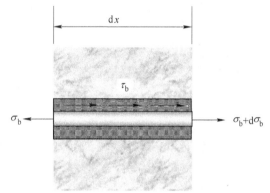

Figure 4-16 Shear stress vs. axial stress

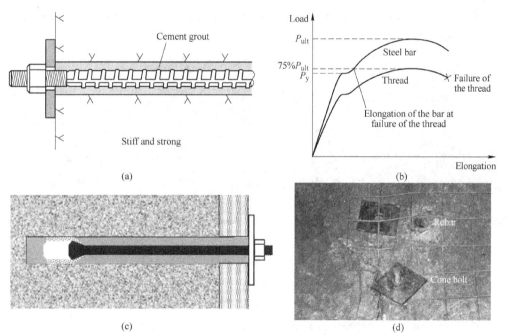

Figure 4-17 A practical problem of rebar bolts
(a) Rebar bolts; (b) Test elongation of the bars; (c) Cone bolts; (d) Field observation

(2) Large rock deformation needs for adaption in highly stress rock →Ductile and strong rock bolts.

4.1.4 Examples for rock support design

(1) Problem description.

1) A mine stope experienced a dramatic change in rock failure and deformation within a short period (as shown in Figure 4-18).

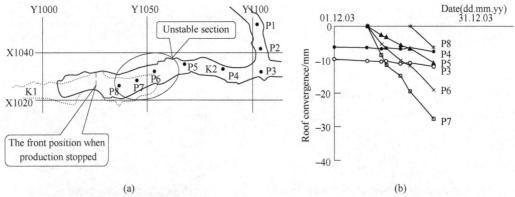

Figure 4-18 Unstable section position and roof convergence

(a) Stope K2 and the unstable section; (b) Roof convergence measured in the stope

2) The stope was closed down because of the instability problem.

3) Rehabilitation was wanted to remove the ore left in the stope.

(2) Rock mechanics assessment.

1) Shear fractures extended to a certain depth in the rock behind the hanging wall surface (as shown in Figure 4-19).

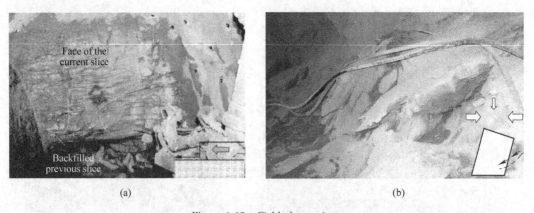

Figure 4-19 Field observations

(a) Surface-parallel fractures behind the roof; (b) Shear fractures in the wall

2) The roof was fractured with surface-parallel fractures.

3) The foot wall was little damaged and could be able to provide a good support to the roof.

Thus: The hanging wall and the roof should be considered as one unit in rock support design.

(3) Support design and result (as shown in Figure 4-20 and Figure 4-21).

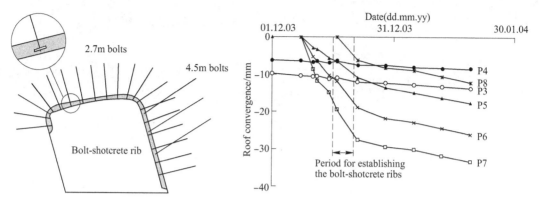

Figure 4-20 Rock support design scheme   Figure 4-21 In-situ observation

4.2 Circular cooperative controlling of the extremely large cross-section roadways with water spraying and water burst in the area of the coal mass and mudstone chamber group

With the large-scale exploiting of coal resources, the number of exploited mine sites and roadways under the complex condition is increasing day by day. In coal mines in China, the quantity of roadways is large and the geological condition in production is complicated. Among them, the horsehead gate in the junction between the shaft and the bottom shaft station belongs to the most typical core "throat" section. The challenge of controlling the surrounding rock mass in this area restricts the development, production and safety of the whole mine site. Additionally, it has an important guidance significance in preventing and solving the mine site rock mass disasters.

4.2.1 Geological production condition of the Xinzhi Coal Mine

The excavating height of the horsehead gate in the junction of the Shangpaoti intake vertical shaft bottom in the Xinzhi Coal Mine is ranged from 4.6m to 7.45m. The net height is ranged from 4.15m to 7m. Additionally, the excavating width is 6.2m and the net width is 5.3m. As for the length, it is 21.5m. The Shangpaoti intake shaft horsehead is tunnelled along the floor mudstone of the coal seam 5#. The roof of the horsehead gate is black mudstones and the thickness of it is 2.9m. Above it, there is a coal seam 5# with a thickness of 1.4m. On the top of the coal seam, there are black mudstones with a thickness of 7.7m and the fine sandstone that is gray and has the thick layer form. The water in the fractures of the kinds of sandy aquifers which are in the roof of the coal seam 5#, during the tunnelling process of the shaft station of the Shangpaoti intake shaft, is the main water sources. The aquosity is relatively weak. It is predicted that during the tunnelling process, there will not be large water bursting. The normal water bursting is $3 \sim 5 m^3/h$ and the maximum water bursting is $15 m^3/h$. The horsehead of the Shangpaoti intake vertical shaft is located in level 310. The position relationship between the auxiliary shaft horsehead and the surrounding chambers is shown in Figure 4-22.

The specific damaging situation of the horsehead gate of the Shangpaoti air shaft is: collapsing of the concrete layer after it is opened. Additionally, rib spalling occurs. Furthermore, collapsing of

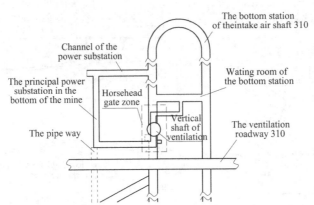

Figure 4-22 The position relationship between the Shangpaoti vertical air shaft and the surrounding chambers

the roof occurs and the depth is ranged from 500mm to 1000mm. Rupture occurs in the lining with different extents. Some linings even collapse. The wall moves towards the internal direction and the displacement magnitude is ranged from 800mm to 1200mm. This leads to the consequence that the adjacent waiting room and electric power substation are squeezed to failure. The decreasing value of the roadway width is 1000mm in average. The phenomenon of roadway floor heaving is serious. At the north side of the horsehead gate, the maximum floor heaving is 2500mm and the average floor heaving is 1750mm, as shown in Figure 4-23.

Figure 4-23 The photos showing the failure status of the horsehead gate of the Shangpaoti air shaft

(a) Failure of the concrete and the roof collapsing; (b) Collapsing failure of the horsehead side;
(c) Splitting of the concrete structure of the horsehead side wall; (d) Failure of the temporary support with the U-steel in the horsehead

4.2.2 The failure mechanism and controlling principle of the roadways with extremely large cross-section in the chamber group

The reserve condition and the production condition of the coal mine roadways are complicated and various. In the practical production, there are kinds of cross-section with different shapes. There are also many different types in the roadway space structure, mainly including the parallel roadway, intersection at the same layer position and the intersection of different layers. For the analysis of the failure mechanism of the roadways with extremely large cross-section in the chamber group, six aspects including the weakness of the surrounding rock masses, the complicated structure of the chamber group, the speciality of the stress field, the high content of the stress concentration, water spraying of the roof and the extremely large cross-section, should be mainly considered.

4.2.2.1 The main reason of the violent failure of the horsehead gate

(1) The roadway surrounding rock masses are relatively weak mudstones and the coal mass bodies are interbedded. The rock mass structure is fractured, leading to the weak strength and the poor stability;

(2) The roadway is located in the junction between the shaft and the station. There are many chambers around the roadway. The space that is located in complicated;

(3) The water spraying and water burst effect of the roadway. The surrounding rock masses are encountered with water. Then, the strength is decreased and the anchoring force of the grout cannot be fully developed;

(4) The roadway cross-section is large and the roadway is complicated. Furthermore, the intersected connecting points are various. The maximum net cross-section is up to $65m^2$;

(5) Due to the effect of the additional stress of the shaft and the concentrated stress of the chamber group, the stress field is extremely complicated;

(6) In the original design of the roadways, the reinforcement type, parameters and structure are not reasonable. Particularly, the controlling on the floor is not considered.

4.2.2.2 Controlling principle of the roadways with extremely large cross-section in the chamber group

Based on the failure mechanism analysis of the roadways with extremely large cross-section in the chamber group area, the stability controlling principle of the roadways with the violent deformation is constructed, as shown in Figure 4-24.

4.2.3 Circular cooperative controlling technology in terms of grouting, anchoring, supporting and casting

4.2.3.1 The circular optimising controlling theory of the roadway reinforcement

(1) The structural mechanical model of the shell support with the "U" shape is shown in Figure 4-25.

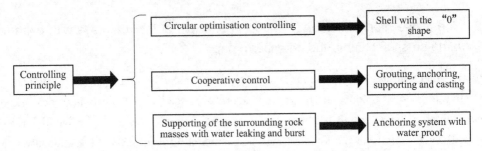

Figure 4-24 Stability controlling principle of the roadways with severe deformation

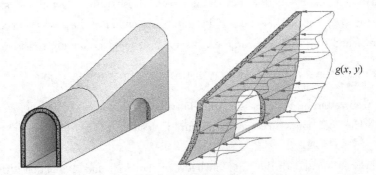

Figure 4-25 The structural mechanical model of the shell support with the "U" shape

Calculating the bending moment that is formed by the loading at a certain position in the roadway side with the following equation:

$$M_{\text{actual}} = \int_0^x \int_0^y g(x,y) y \mathrm{d}x \mathrm{d}y \qquad (4\text{-}2)$$

(2) The structural mechanical model of the shell support with the "0" shape is shown in Figure 4-26.

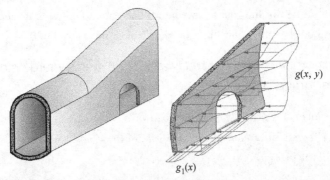

Figure 4-26 The structural mechanical model of the shell support with the "0" shape

According to the calculating method of the roadway side bending moment bearing capacity, the roadway side bearing bending moment of the shell support with "0" shape can be acquired:

$$M_{0\text{shell}} = \int_0^x \int_0^y g(x,y) y \mathrm{d}x \mathrm{d}y - M_0 \qquad (4\text{-}3)$$

Comparing and analysing the structure mechanical model and its roadway side bearing moment calculation equation of the "U" shape shell and the "0" shape shell, the "U" shape shell supporting type has more apparent superiority, compared with the "0" shape shell.

4.2.3.2 The cooperative controlling system in terms of grouting, anchoring, supporting and casting

Through the in-situ study and surveying, it can be known that to realise the effective processing of the roadways, the corresponding measures on the roadway roof and two sides should not only be conducted. The processing of the floor should but also be equally significant. Therefore, the processing scheme should consider the roadway roof, two sides and the floor at the same time. The joint processing method of the roadway with extremely large cross-section in the chamber grout where the coal and mudstones are interbedded, is the cooperative controlling technology in terms of grouting, anchoring, supporting and casting, in which the grouting reinforcing technology in the shallow and deep boreholes, the circular integrated reinforcement technology with rock bolts and cable bolts, and the combined casting technology with concrete and U-steel, are integrated together. This is shown in Figure 4-27. Its principle is:

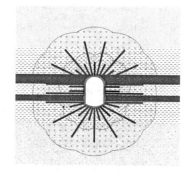

Figure 4-27 The cooperative controlling principle diagram of the grouting, anchoring, supporting and casting

(1) The grouting in deep boreholes and the rock bolt (cable bolt) anchoring effect are acted on the deep surrounding rock masses in the roadway. This prevents the developing of the loose failure of the fractured surrounding rock masses into the deep area. And the deep reinforcing arch with the "0" shape is generated.

(2) The rock bolts (cable bolts) along the full cross-section and the grouting layer in the shallow borehole are acted on the shallow surrounding rock masses of the roadway. Therefore, the whole strength of the surrounding rock masses and the bearing capacity are enhanced. The middle reinforcing arch with the "0" shape is formed.

(3) The U-steel support and the high strength steel reinforced concrete casting are acted on the surrounding rock mass surface of the roadway. This is to improve the rigid supporting body, forming the reinforcing arch with the "0" shape at the surface.

The circular cooperative controlling effectively improves the surrounding rock mass strength. This increases the supporting resistance and its acting range. This effectively improves the equivalent thickness and the bearing range of the surrounding rock masses with the "0" shape. The surface plastic range of the roadway is decreased with large extent, which is beneficial for keeping the roadway stability. The superiority of the cooperative controlling with grouting, anchoring, supporting and casting is mainly reflected in the following aspects:

(1) First, through grouting, the rock mass strength and the construction safety can be increased. Then, from the deep area of the surrounding rock masses to the shallow area, the rock

bolt reinforcement, the cable bolt reinforcement and the cast technique are implemented. This realises the proper optimising of the controlling principle of the roadway surrounding rock masses, the techniques and the effects.

(2) In the roadway surrounding rock masses, the reinforced arch is formed. This makes the internal weak and fractured coal masses and rock masses be subjected to the radial constraint and shearing constraint. This makes the surrounding rock masses be subjected to the three-dimensional stress state. Then, the self-bearing capacity of the weak and fractured coal masses and rock masses is improved.

(3) Rock bolts and cable bolts are respectively acted on the deep area of the roadway surrounding rock masses and the middle circular reinforced arch. Then, the pre-tension and the controlling are realised on the deep and shallow surrounding rock mases. Consequently, the purpose of the cooperative bearing from the deep area to the shallow area is realised.

(4) In the cooperative controlling system, the deep reinforced arch and the central reinforced arch are connected together through cable bolts. The central reinforced arch and the surface reinforced arch are connected through grouted rock bolts. This avoids the dispersive bearing of the reinforced arch for each layer.

4.2.4 The circular cooperative controlling technical scheme in terms of grouting, anchoring, supporting and casting

According to the above circular cooperative controlling principle and technology in terms of grouting, anchoring, supporting and casting, meanwhile based on the numerical simulation, mechanical analysis and the engineering comparison, it is comprehensively determined that the circular cooperative controlling technical scheme in terms of grouting, anchoring, supporting and casting for the horsehead gate of the Xinzhi Coal Mine is shown in Figure 4-28.

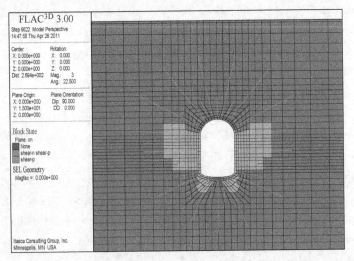

Figure 4-28 The numerical calculating results with $FLAC^{3D}$ based on the circular cooperative controlling in terms of grouting, anchoring, supporting and casting

4.2.4.1 Grouted reinforcing with different times in the deep-shallow boreholes

(1) For the grouted rock bolts in the deep borehole, the hollow grouted rock bolts with left lateral threaded rods with the diameter of 25mm is selected. The rock bolt length is 4000mm and the diameter of the drilled borehole is ranged from 38mm to 43mm. The depth of the drilled borehole is 5000mm. The diameter of the rock bolt hollow hole is more than 17mm. The rock bolt spacing is 1.6m. The top rock bolt is located in the middle of the circular arch of the roadway. As for the other rock bolts, they are set along two sides with an interval of 1.6m. For the rock bolt angle, the grouted rock bolts on the semi-circular arch are perpendicular to the rock mass surface. As for the grouted rock bolts installed in vertical walls of two sides, except the lowest rock bolts, they are all perpendicular with the roadway surface. In the vertical wall, the lowest grouted rock bolts have a downward intersection angle of 15 degrees with the horizontal face.

(2) For the second grouting in the shallow borehole behind the wall, the hollow grouted rock bolts with the threaded rock bolt rod and diameter of 25mm are selected. The rock bolt length is 2200mm and the drilled borehole diameter is ranged from 38mm to 43mm. The depth of the drilled borehole is 2500mm. The diameter of the rock bolt central hole is not less than 17mm and the grouted rock bolt spacing is 1.4m. The top rock bolt is located in the middle of the circular arch of the roadway. As for the rock bolts, they are set along two sides with the interval of 1.4m. The rock bolt angle is consistent with the grouted rock bolt angle in the deep borehole. Before the concrete is cast, the grouted rock bolts in the shallow boreholes are installed. First, the grouting borehole is prepared. After the concrete is poured and reaches a certain strength, the second grouting is conducted.

4.2.4.2 The circular integrated reinforcement type and parameters in the weak roadways

Before the horsehead gate is reinforced with rock bolts and cable bolts, first, the original roadway reinforcement structure should be eliminated and expanded. Then, the combined reinforcement which is composed of rock bolts and single cable bolt is conducted on the roof side and the floor.

A The rock bolt reinforcement

For the roof side rock bolt, the left ribbed threaded high strength rock bolts without longitudinal ribs ($\phi 22mm \times 3000mm$) are used. The anchorage length is 1760mm. The rock bolt is installed perpendicular with the roadway cross-section. For the roadway side, the lowest rock bolt is installed with a downward inclination angle of 15 degrees. For the rock bolt installation, a rock bolt is installed along the middle line of the roadway roof. The rock bolt interval is 700mm and the spacing is 700mm. For the floor rock bolt, the right ribbed rock bolts without longitudinal ribs ($\phi 20mm \times 2000mm$) are used. The anchoring length is 880mm. The rock bolt is installed perpendicular with the floor cross-section. As for the rock bolt installation, a rock bolt is installed along the middle line of the roadway floor. The rock bolt interval is 700mm and the spacing is 700mm.

B The single cable bolt

For the roof, the high strength pre-tensioned strand ($\phi 21.6$mm) is used. The depth of the cable bolt hole is 10000mm and the cable bolt length is 10300mm. The diameter of the drilled borehole is 28mm and the anchoring length is 2640mm. For the single cable bolt, they are installed in the middle of two rows of rock bolts. And they are installed from the central line of the roadway for 700mm. The cable bolt interval is 1400mm and the spacing is 1400mm. For the roadway side, the high strength pretensioned strand ($\phi 21.6$mm) is used. The depth of the cable bolt hole is 8000mm and the cable bolt length is 8300mm. The diameter of the drilled borehole is 28mm and the anchoring length is 2640mm. The cable bolt arrangement is in the same row with the roof cable bolts. The top cable bolt at the roadway side has a distance of 1400mm to the lowest cable bolt in the roof. As for the other spacing, it is 1400mm. The cable bolt row spacing at the roadway side is 1400mm.

4.2.4.3 The parameters of the U-steel and reinforced concrete casting in the roadway with extremely large cross-section

The material of the U-steel is the steel 36 U with the line density of 36kg/m. Its position is located in the middle between two layers of steels in the concrete supporting. For the strength level of the reinforced concrete, the commercial high strength concrete of C50 is used. For the concrete casting thickness design, in the whole cross-section, the cast thickness is 600mm.

Overall, the specific technical scheme and parameters of the horsehead gate processing is shown in Figure 4-29.

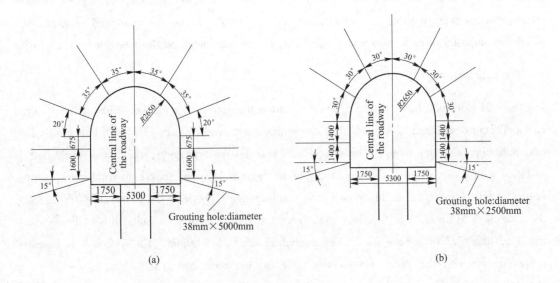

(a)　　　(b)

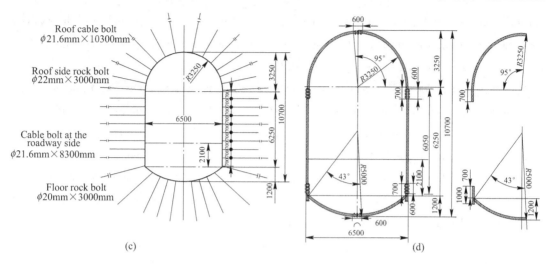

Figure 4-29 The circular cooperative controlling technical scheme in terms of grouting, anchoring, supporting and casting

(a) Typical cross-section of the grouting in the deep borehole; (b) Typical cross-section of the grouting in the shallow borehole; (c) The highest cross-section diagram of the reinforcement with rock bolts and cable bolts; (d) Connection diagram of the U-steel support and the design of the standard component

4.2.5 In-situ practice and observation

The specific construction process of the horsehead gate processing is shown in Figure 4-30.

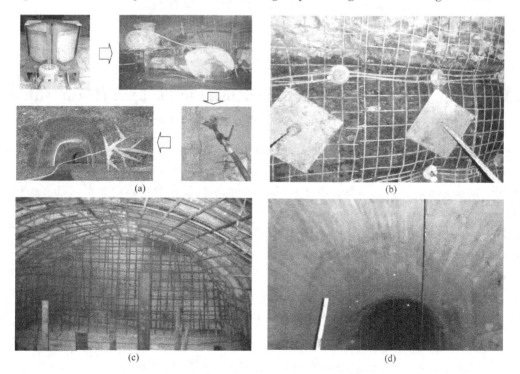

Figure 4-30 The processing diagram of the horsehead gate construction

(a) Construction of the roadway grouting; (b) Completion of the rock bolt and cable bolt construction; (c) Construction of the U-steel support; (d) Completion of the casting construction

It can be divided into the following steps: grouting in the deep borehole → expanding of the roadway → cement grouting to maintain the roadway side → meshing and cable bolting → U-steel shed → concrete casting → grouting in the shallow borehole.

After the construction of the roadway is completed, three measuring points are installed in the roadway to monitor the roof subsidence of the roof and the deformation values of two sides. The purpose of it is to determine the roadway supporting and reinforcing effect. The observation results, as shown in Figure 4-31, show that for the extremely large cross-section roadway, no roof subsidence and deformation of two sides occur. Additionally, the macroscopic cracks and fractures which are resulted by loading does not occur. Therefore, it meets the requirement of the reinforcement design and acquires favourable reinforcing effect.

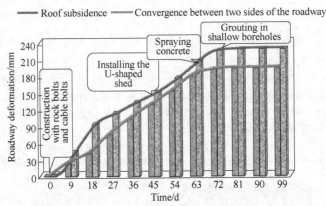

Figure 4-31 The surface displacement observing curve at the typical observation station

4.3 The cable truss combined reinforcement in the open-off cut with the composite mudstone roof and large mining height

The rock bolt reinforcement technologies in coal mines in China promote the advance of the controlling technologies in the controlling of the roadway surrounding rock mass. Additionally, the remarkable economic and social benefits have been acquired. Nevertheless, the damaging and roof collapsing of the large span coal roadway which is composed of mudstones seriously make a threat on the personal safety of the coal operators. Furthermore, these problems also have an effect on the ventilation, gas, transportation and pedestrian, restraining the safe and efficient production of the mine sites. Using the traditional rock bolt (cable bolt) reinforcement technologies is difficult to realise the effective controlling of the hydraulic support-surrounding rock mass system. Realising the effective controlling of the large span coal roadways which is composed of mudstones has already become the urgent demanding of the mining industry.

4.3.1 Geological production condition of the Bailong Coal Mine

The fully mechanised working face with large mining height 2-1101 in the Bailong Coal Mine mainly exploits the coal seams 1# and 2#. The mean coal seam dip angle is 12 degrees. For the coal seam 1#, the mean thickness is 2.2m. As for the coal seam 2#, the mean thickness is 2.1m. Between coal seams 1# and 2#, there is a mudstone interlayer with the thickness ranging from 0.1m to 1.0m. For the immediate floor of the coal seam, it is the composite roof which is composed of 6 layers of mudstones and the coal seam line. The average thickness is 2.2m. Above the composite floor, it is the weak rock strata with the thickness ranging from 4m to 6m (mudstones or the sandy sandstones). The open-off cut of the working face 2-1101 is along the east-west direction. The designed length is 150m and the cross-section is rectangular. The net width is 7.0m and the height is 3.2m. The open-off cut is tunnelled in the roof along the coal seam 1#. The roof belongs to the composite fractured roof with the typical water spraying. The coal and rock bar chart of the working face is shown in Figure 4-32.

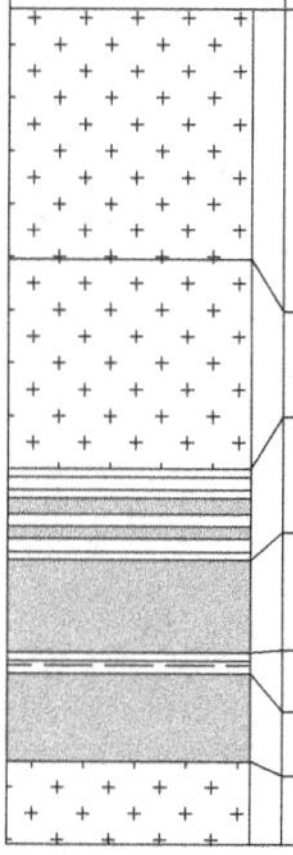

Column profile	Rock name	Thickness /m	Lithologic character
	Sandstone with medium size	7~8	Sandstone K_8 with medium particle size. Mainly composed of quartz and then feldspar. Inclined beddings. Fractures of specular coal and be found in the local area. Calcite cementation separation is favourable
	Mudstone or the muddy sandstone	4~6	Grey fine sandstone and block size
	Mudstone or the coal line	2.2	Grey mudstone. Interactive reserve between 6 layers of mudstones and coal line
	Coal 1#	2.15~2.30 / 2.2	Mainly composed of semi-bright coal and bright coal. Between them, there are dark coal seam with thin layer
	Mudstone	0.1~1.0	Ash black. Block mass, and muddy structure
	Coal 2#	2.05~2.20 / 2.10	Brown-black. The layer is relatively stable
	Fine sandstone	2.0	Grey fine sandstone. The black mudstone stripe is mixed

Figure 4-32 The coal and rock bar graph

4.3.2 Rock bolt and cable bolt problems in the mudstone coal roadways with large span

The rock bolt reinforcement technologies in coal mines in China promote the scientific advance of the controlling in the surrounding rock mass of roadways. Additionally, the remarkable economic and social benefits have been acquired. However, for the mudstone coal roadways with large span, using the traditional rock bolt and cable bolt reinforcement technology is difficult to effectively control the hydraulic support-surround rock mass system. The reinforcement problems of the mudstone coal roadway with large span are reflected in the following six aspects:

(1) The rock mass strength of the mudstone is low. The bedded fractures are developed. It is easy to generate the bed separation failure and collapsing;

(2) The coal mass is loose and weak. The joints and fractures are developed. The load on the coal side of the roadway with large span is increases with times. The coal side failure is increases;

(3) For the large span roadway, it is easy to form the tensile stress in the middle of the roof and the shear stress concentration in the end. It is not beneficial for the roof maintaining;

(4) The increasing of the roadway span makes the stress and deformation in the mudstone roof increase dramatically in the square order and the cubic order;

(5) The fracturing of the mudstone and coal mass in the roadway with large cross-section has the large range. The anchorage force of the original rock bolts and cable bolts cannot be guaranteed;

(6) The anchorage point of the single cable bolt is located in the above weak rock strata. It is easy to fail with the severe sinking of the roof and the bed separation.

4.3.3 The high pretension cable bolt truss controlling system

4.3.3.1 The structure of the high pretension cable truss

The pretension cable truss controlling system is the highly reliable cable truss structure with multiple active reinforcement. In this structure, the newly developed special cable truss connecting instrument is regarded as the core component. Specifically, its joint reinforcement system is composed of several parts, namely the high strength cable bolt, the locking instrument, the special connecting instrument and the anchorage grout. The reinforcement principle of the high pretension cable truss system is shown in Figure 4-33.

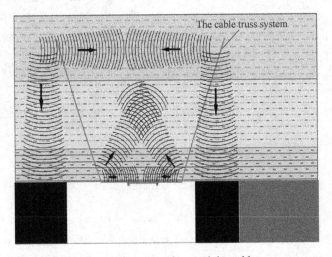

Figure 4-33 The reinforcement theory of the cable truss system

The high pretension cable truss system regards the rock masses which are subjected to the compressive state and are located in the deep area of the roadway two shoulders as the anchorage point and the basement of the load bearing structure. Specifically, the high pretensioned force is used to tighten and lock tightly two steel strands. It directly acts on the shallow rock masses of the immediate roof, providing the horizontal pretension force to improve the stress state of the roof. The mechanical properties of the rock masses in the shallow position can be enhanced and the anti-deforming performance of the rock masses can be improved. It controls the inconsistent deformation of the bedded roof. The cable truss is the active reinforcement structure that can provide the squeezing and compressing stresses in the roadway roof along the horizontal direction and the vertical direction at the same time. Consequently, the coal mass and the rock mass in the anchorage area are subjected to the squeezing and compressing state along the vertical direction and the horizontal direction. The active reinforcement force that is generated by the pretension force of the cable truss system makes the roadway roof have the moving tendency along the upward direction. Therefore, this makes part of the roof subsidence be counteracted. In the bending deformation process of the roadway roof, the tensile stress in which the cable bolt suffered increases. Consequently, the squeezing and compressing force that the coal mass and the rock mass suffered in the anchorage area increases accordingly.

4.3.3.2 The superiority of the high pretension cable truss structure

Compared with the traditional normal rock bolt (cable bolt) reinforcement method, the reinforcement mechanism of the pretension cable truss has fundamental changing. Its superiority is reflected in the following aspects:

(1) The cable truss system can provide the active reinforcement force along the horizontal direction and the vertical direction at the same time. Furthermore, the tensile stress that the cable truss system suffers and the reinforcement force that the cable truss system provides increase with the roof deformation. This system effectively decreases the maximum tensile stress in the coal mass and rock masses in the middle area of the roadway. This is beneficial for the coal mass and rock mass being subjected to multiple compressive stress state. Furthermore, this can provide the strength of the coal masses and rock masses. Also, this can improve the anti-deformation failure performance of coal masses and rock masses.

(2) The length of the cable bolt in the cable truss system is long and the anti-shearing performance is strong. It obliquely crosses the area where the maximal shear stress occurs in the roof above the coal pillar. Furthermore, the applied range is large. It jointly supports the shear stress with the corner rock bolts and the roof. And it can effectively control the shear failure of the roof.

(3) In the cable truss system, there is a line contact between the strand and the roof. The load in the strand can be transferred continuously and it is convenient to apply large pretension. The reinforcement range is large. The suffered load of the loose and fractured roof is favourable.

(4) The anchorage point of the cable truss is located in the deep rock masses in the shoulder that is subjected to three-dimensional stress condition. It is not easy to be influenced by the roof bed separation and deformation. This provides reliable and stable load bearing basement for the cable truss system to develop high anchorage force.

(5) During the roof bending and sinking process, the anchorage point of the two sides of the cable truss structure shift inwards. The increment of the suffered load is relatively slow. The reinforcement structure is not easy to fail. The interlocking structure can control the further deformation of the roof and prevent the severe roof accident.

In conclusion, after the cable truss system is used to reinforce the roof in the roadway, the maximum tensile stress and the maximum shear stress that the rock strata bear in the anchorage area decrease. Additionally, the range that the tensile stress affects decreases. Furthermore, the deflection decreases and the structural property is enhanced. Consequently, compared with the normal rock bolt-cable bolt reinforcement, the stability of the rock strata in the anchorage area is largely improved, as shown in Figure 4-34.

Figure 4-34   The comparison diagram showing the high pretension cable truss and the single cable reinforcement
(a) The joint controlling of the high pretension cable truss structure; (b) The traditional single cable reinforcement

4.3.3.3   The joint controlling theory of the cable truss

According to the function relationship between the cable truss and the surrounding rock masses, the mechanical model as shown in Figure 4-35 can be established.

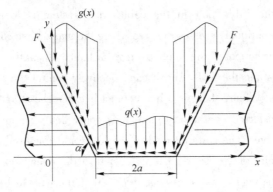

Figure 4-35   The mechanical model of the high pretension cable truss system

According to the mechanical model, it can be calculated that the pretension force of the cable truss is:

$$F' = \frac{b(k_1 + k_2)(\lambda + \cot\alpha)\cos\alpha + 2a(f_2 + \cot\alpha)}{2}\gamma h \qquad (4\text{-}4)$$

The tensile strength of the cable bolt is:

$$\sigma_{ten} > \frac{2a + b(k_1 + k_2)[1 + f_1(1 + \lambda\tan\alpha)\sin\alpha]\cos\alpha}{2s_{cable}\sin\alpha}\gamma h \qquad (4\text{-}5)$$

Where, k_1, k_2 are the coefficients of the linear equation $g(x)$; λ is the lateral coefficient.

4.3.4 The composite reinforcement scheme of the cable truss system in the open-off cut

4.3.4.1 The reinforcement scheme of the coal open-off cut 2-1101 in the Bailong Coal Mine

According to the practical geological production condition of the coal open-off cut 2-1102 in the Bailong Coal Mine, the mechanical analysis, the orthogonal experiment and the numerical calculation are used together to determine the reinforcement scheme of the coal open-off cut 2-1101, as shown in Figure 4-36.

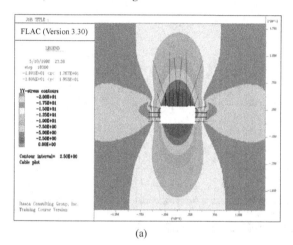
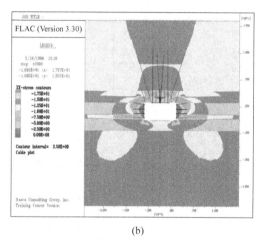

(a) (b)

Figure 4-36 The numerical simulation calculation results of the reinforcement scheme of the coal open-off cut
(a) Vertical stress distribution diagram of the roadway surrounding rock masses;
(b) Horizontal stress distribution diagram of the roadway surrounding rock masses

The specific layout and the reinforcement form of the fully mechanised working face with large mining height 2-1101 are: the coal open-off cut is formed by two times. First, along the permanent support side, the excavation is conducted. For the rectangular cross-section, the width is 3.5m and the height is 3.2m. Second, along the coal wall side of the fully mechanised working face, the excavation is conducted. For the rectangular cross-section, the width is 3.5m and the height is 3.2m. The new high pretension cable truss is matched with the single pretension cable bolt, the high strength pretension rock bolt, the W-shaped steel bet and the metal mesh. Then, this composite support is used to strengthen the mudstone roof of the coal roadway. The high strength pretension rock bolt, the steel bearing beam and the metal mesh are combined to strengthen two sides of the roadway. This make the weak surrounding rock masses of the roadway form an integrated entity. Then, it can effectively control the roadway deformation, as shown in Figure 4-37.

A The first horizontal support form

For the roof support, the rock bolt is the left lateral threaded high strength rock bolts without longitudinal ribs 20#. The length of the rock bolt is 2.5m. The spacing of the rock bolt is 0.8m and 5 rock bolts are installed in each row. The interval is 0.8m. Along the permanent support coal

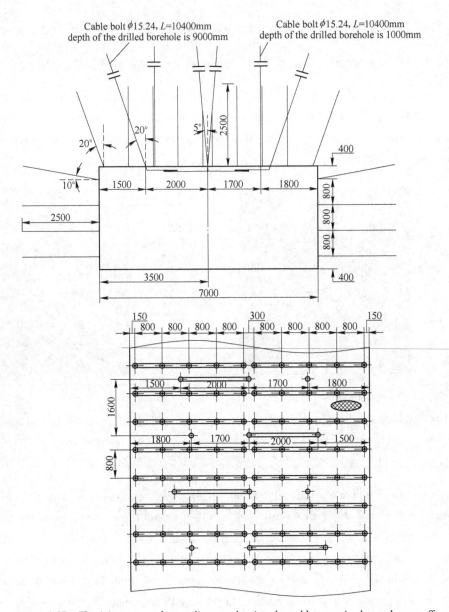

Figure 4-37 The joint support layout diagram showing the cable truss in the coal open-off cut at the fully mechanised working face 2-1101

side, the corner rock bolt of the roof has a distance of 150mm to the coal side. As for the corner rock bolt of the roof along the temporary support coal side, the distance between it and the coal side is 150mm. The single cable material is the high strength loose pretension strand (1×7) with a diameter of 15.24mm. The length of the cable bolt is 10.4m. Every four rows of 3.2m, a single cable bolt is installed. The cable bolt is installed in the position where the distance from it to the permanent support coal side is 1.8m in the roof. The material of the cable bolt in the cable truss system is the high strength and loose pretension steel strand (1×7) with a diameter of 15.24mm. The length of the cable bolt is 10.4m. Every four rows of 3.2m, a set of the cable truss system is installed between the single cable bolt. The bottom span of the cable truss system is 2.0m. The distance between the cable truss that is along the permanent support coal side and the permanent support coal side is 1.5m. The distance between the cable truss that is along the temporary support coal side and the temporary support coal side is 0.

For the roadway side support, the rock bolt is the left lateral threaded high strength rock bolts 20# without longitudinal ribs. The length of the rock bolt is 2.5m. The rock bolt spacing is 0.8m and the interval is 0.8m. The installation angle between the roadway side rock bolt along the roof and the horizontal surface is +10 degrees. As for the rest, they are installed horizontally. Along the entity coal wall side, the point column method is used to provide support. The point column spacing is 0.8m. Every two point columns, 4 wooden backboards are used to bear the side.

B The second horizontal support form

For the roof support, the rock bolt is the left lateral threaded high strength rock bolts 20# without longitudinal ribs. The length is 2.5m. The rock bolt spacing is 0.8m. In each row, there are 5 rock bolts and the spacing is 0.8m. The distance between the roof corner rock bolt that is along the fully mechanised working face coal side and the coal side is 150mm. The distance between the roof corner rock bolt that is located in the centre of the second coal open-off cut and the roof corner rock bolt that is located in the centre of the first coal open-off cut is 300mm. The material of the single cable bolt is the high strength and loose pretension steel strand(1×7) with the diameter of 15.24mm. The length of it is 10.4m. Every four rows of 3.2m, a single cable bolt is installed. The rock bolt is installed in the roof where the distance between it and the coal side of the fully mechanised working face is 1.8m. In the cable truss system, the cable bolt material is the high strength and loose pretension steel strand(1×7) with the diameter of 15.24mm. The length of it is 10.4m. The length of the drilled borehole is 9.0m. Every four rows of 3.2m, a set of the cable truss system is installed between the single cable bolt. The bottom distance of the cable truss is 2.0m. The distance between the cable truss that is along the coal side of the fully mechanised working face and the coal side of the fully mechanised working face is 1.5m. In the cable truss system, the other cable bolt is located in the middle of the coal open-off cut. As for the support form of the coal wall side along the fully mechanised working face, it is same as the support form of the permanent support coal side.

4.3.4.2 Analysis of the ground pressure observation results

The ground pressure observation results indicate that the convergence rate of the roadway cross-section is pretty small. The maximum convergence of two sides is not over than 145mm. The maximum roof sinking is not over than 166mm. The controlling effect of the roadway surrounding rock masses is favourable. For the open-off cut, the convergence velocity in the initial period is larger than that in the later period. For two sides of the roadway and the floor, after around ten days, the convergence velocity of the roadway is to be stable. For the roof, the required time is relatively longer. After around twenty days, it is to be stable. The roof bed separation observation results indicate that the loose bed separation in the top is pretty small. The roof is in the stable state. The rock bolt anchorage force and the pretension moment detecting results indicate that the percent of pass is high. The construction quality of the rock bolt reinforcement in the coal open-off cut 2-1101 is favourable. The reinforcement system works reliably.

4.4 The asymmetric failure mechanism and controlling of the surrounding rock masses in the gob-side tunnelling in the fully mechanised caving area with large cross-section and intense mining

The fully mechanised top coal caving mining, especially the large scale and integrated fully mechanised caving working face with the length ranging from 230m to 300m, is the important development direction of the thick coal seam exploiting with high production and safety. However, the corresponding large scale and high strength fully mechanised caving roadways are subjected to a series of the surrounding rock mass controlling issues such as the large cross-section and the intense mining.

4.4.1 The geological production condition of the Wangjialing Coal Mine

The area 201 in the Wangjialing Coal Mine uses the top coal caving method to exploit the coal seam 2#. The reserves of the coal seam are stable. The average thickness is 6.21m. The mean dip angle is 3 degrees. The immediate roof is the sandy mudstone with the thickness of 2.0m. As for the main roof, it is the fine sandstone with the thickness of 9.6m. For the floor, it is the mudstone with the thickness of 1.61m. The fully mechanised caving working face 20103 is located on the west side of the central roadway in the mining zone 201. Along its north side, it is the gob area of the fully mechanised caving working face 20105. The width of the coal pillar is 8m. Along the north side, it is the entity coal. For the fully mechanised caving working face 20103, the length along the dip direction is 261m and the length along the strike direction is 1490m. The haulage roadway of 20103 has the rectangular cross-section. The specification is 5600mm×3550mm (width × height). Due to the fact that the connection between the mining and tunnelling is tight in the Wangjialing Coal Mine, the haulage roadway of 20103 will be subjected to the influence of the multiple mining. The arrangement form of the roadway in the zone 201 is shown in Figure 4-38.

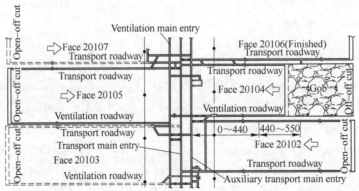

Figure 4-38 The roadway arrangement diagram in the mining zone

In the Wangjialing Coal Mine, the other extremely large-scale ventilation roadway in the fully mechanised caving zone showed the support system failure, large deformation and the severe roof collapsing accident. The length of the roof collapsing is up to 61m and the collapsed height is up to 5m. This leads to the consequence that the roadway is failed for a long time and the traffic is terminated in it. The scientific technical problems such as the roof collapsing in the gob-side roadway in the fully mechanised caving area, has already seriously restricted the safety production of the mine site, as shown in Figure 4-39.

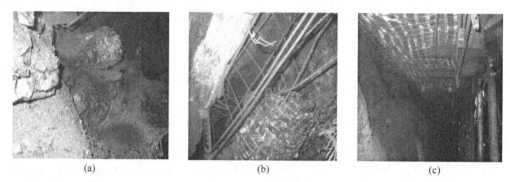

Figure 4-39 The roof and side failure diagram showing the ventilation roadway along the gob area in the fully mechanised caving zone
(a) Collapsing of the side; (b) Collapsing and leaking of the corner of the roof side;
(c) Repairing and reinforcement of the roof collapsing for the roof

4.4.2 The controlling problem of the surrounding rock masses in the fully mechanised caving coal roadway with large cross-section and intense mining

According to the above mentioned practical geological production situation of the Wangjialing Coal Mine, it is acquired that the surrounding rock mass controlling problems of the fully mechanised caving coal roadway with large cross-section and intense mining are as following:

(1) The large scale and high strength fully mechanised caving mining will absolutely lead to the high abutment pressure and the large range. The influence is severe.

(2) The haulage roadway in the large scale fully caving working face requires the functions of belt conveyor, the railway, the ventilation and the pedestrian. Therefore, the requirement on the roadway cross-section is high.

(3) The span increasing leads to the consequence that the roof stress and deformation increase in the square order and in the cubic order respectively. This leads to the middle splitting of the roof and the shear failure of the roadway side corner.

(4) The coal mass is loose. The joints and fractures are developed. The load in the coal side in the roadway with the large cross-section increases with times. The coal side is seriously bloat.

(5) For the coal roadway with the intense mining (the coal pillar width is 19.4m), the large area failure, deformation and the severe roof collapsing accidents of the support system have already occurred.

(6) The width of the coal pillar is 8m. The surrounding rock masses and the stress environment of the roadway deteriorate. The difficulty in the surrounding rock masses controlling further increases.

(7) For the ground pressure around the roadway, the new characters that the horizontal compressing and the vertical sinking movement occurred meanwhile. Furthermore, the deformation of the surrounding rock masses shows asymmetric.

(8) The traditional rock bolt and cable bolt reinforcement cannot adopt to this new law. Therefore, new challenge is proposed for the roadway surrounding rock mass controlling theory and technologies.

4.4.3 The asymmetric failure mechanism of the coal and rock masses in the roof of the gob-side coal roadway in the fully mechanised caving

Due to the character structure of the surrounding rock masses of the coal roadway along the fully mechanised gob and the difference on the mining influence extent along the horizontal direction along the cross-section of the roadway, the relative large asymmetric character of the ground pressure appearance of the roof along the vertical direction and horizontal direction along the centre axis of the roadway cross section occurs. Furthermore, this asymmetric character is more serious and prominent under the condition that the shallow coal pillars are used to protect the roadway. Therefore, the research regarding the asymmetric failure mechanism of the coal masses and rock masses of the roadway along the fully mechanised caving gob, such as the structural character of the overlying rock strata of the coal roadway under the intense mining, and the movement form of the overlying rock strata under the mining influence, has important theory and practical significance.

4.4.3.1 The structural character of the overlying rock strata of the intense mining coal roadway

For the previous section adjacent fully mechanised caving working face, when the forward advancing is conducted to a certain distance from the open-off cut, the initial weighting of the roof forms the rupture from of "O-X". It means that the paralleled rupture lines I_1 and I_2 are initially formed in the centre of the exposed main roof and two sides that have relatively longer length. Then, the rupture line II is formed along the side that has the relatively shorter length. Furthermore, they are connected with the rupture lines I_1 and I_2. Finally, the rock strata in the main roof rotate along the rupture lines I and II. Then, the rupture line III of separated blocks is formed. Consequently, the structural blocks 1 and 2 are formed. As the working face continues to advance, periodic collapsing occurs in the roof. Consequently, the rupture line I_2 occurs in proper order. Furthermore, it rotates back along the surrounding rupture line II and the periodic roof collapsing is formed. Meanwhile, the new structural blocks 1 and 3 are formed, as shown in Figure 4-40.

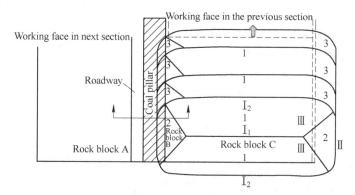

Figure 4-40 The structural relationship of the overlying rock strata of the mining roadway

According to the calculation of the elastic and plastic ultimate equilibrium theory, the distance between the rupture position and the coal face along the gob-side in the previous section can be calculated with the following equation:

$$x_0 = \frac{\lambda m}{2\tan\varphi} \ln\left(\frac{k\gamma H + \dfrac{C}{\tan\varphi}}{\dfrac{C}{\tan\varphi} + \dfrac{P}{\lambda}}\right) \quad (4\text{-}6)$$

Where, x_0 is the interval between the rupture line of the core block and the coal face, m; C is the cohesive force of the coal mass, MPa; φ is the internal friction angle of the coal mass, (°); P is the supporting resistance along the coal pillar side, MPa; m is the average mining thickness of the district, m; λ is the coefficient of the lateral pressure; k is the maximum stress concentration coefficient; γ is the average volume weight of the overlying strata, kN/m^3; H is the average buried depth of the fully mechanised working face in the mining district, m.

4.4.3.2 The mechanical model of the overlying rock strata for the fully mechanised caving coal roadway under the intense mining condition

After the retreat mining is conducted on the adjacent fully mechanised caving working face, at the connection of it along the lateral direction and the next working face, rupture occurred in the roof above the coal pillar. And the arc triangular block is generated. After rupture, the triangular block rotates back and sinks. After one end of the rock block rotates, it contacts the gangue in the gob area. Although the rock block has a certain back rotation and sinking, it interlocks with the rock block and rock masses, forming the "large structure of the voussoir beam". Then, the lateral abutment pressure is generated on the roadway roof. Based on this, the entire mechanical model regarding the overlying rock strata above the fully mechanised caving roadway with large cross-section and intense mining is established, as shown in Figure 4-41.

4.4.3.3 The movement form of the overlying rock strata of the coal roadway under the intense mining condition and the displacement variation law

The software of UDEC is used to establish the movement model of the overlying rock strata under

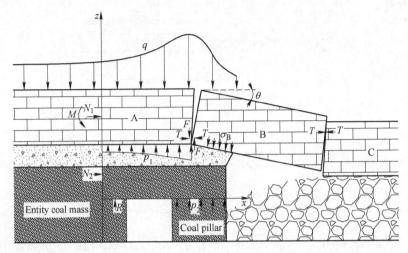

Figure 4-41 The mechanical model of the overlying rock strata for the fully mechanised caving roadway with large cross-section and intense mining

the intense mining condition. The width of the coal pillar is 8m. The length of the model along the strike direction is 200m. The vertical height is 60m. The roadway dimension is 5.6m × 4.0m (width × height). For the rock block, the Mohr-coulomb model is used. As for the joint, the contact Coulomb slippage model is used.

A The movement form of the overlying rock strata for the fully mechanised coal roadway with large cross-section and the intense mining

The UDEC is used to calculate the model, to analyse the movement form, the plastic failure and the movement law character of the large structure of the roof above the roadway along the gob area when the timestep is different, as shown in Figure 4-42. The results indicate that the core block B has the back-rotation movement, which is the core reason leading to the asymmetric deformation

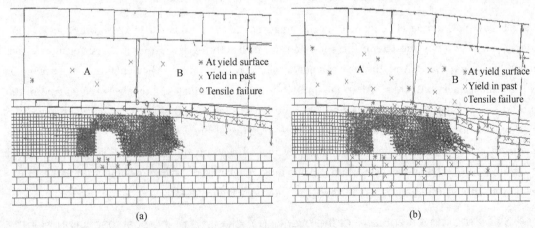

Figure 4-42 The surrounding rock mass displacement field and the failure field variation of the gob-side roadway
(a) 4000step; (b) 6000step

and the horizontal shearing movement of the roof above the gob-side roadway. However, the instability failure of the small structure of the gob-side roadway will further promotes the back-rotation and slippage deformation of the large structure. Furthermore, this will lead to the second rupture of the core block in the deep area. Guaranteeing that the large structure and the small structure of the surrounding rock masses have the double stability, making realisation of the stability of the surrounding rock masses of the gob-side roadways.

B The displacement field variation character of the fully mechanised caving coal roadway with large cross-section and intense mining

As show in Figure 4-43, during the tunnelling period, the vertical displacement of the roof between the entity coal side and the coal pillar, shows the inclined one-character form. In the range of 2.0m, the roof sinking along the coal pillar side is larger than the entity coal side. The maximum subsidence is located in the roadway centre along the coal pillar side for 200mm to 600mm. During the retreat mining period, in the deep area (>3.0m), the vertical displacement from the entity coal side to the coal pillar side increases for approximately 54mm. In the shallow range of 3.0m, the subsidence of the surrounding rock masses increases for approximately 100mm. However, the entire deformation still shows the asymmetric character, showing the large deformation with the form of "loosing and expanding".

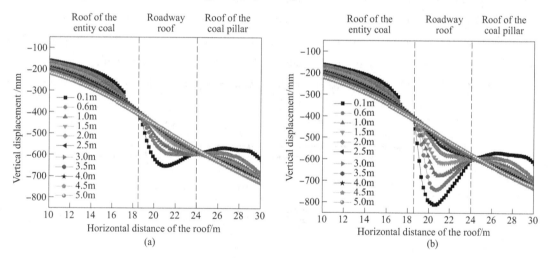

Figure 4-43 The displacement variation character of the fully
mechanised caving roadway with large cross-section and intense mining
(a) The vertical displacement distribution character in the tunneling period;
(b) The vertical displacement distribution character in the retreat mining period

4.4.4 The controlling system of the newly developed high pretension cable truss system

The traditional rock bolt (cable bolt) reinforcement technology cannot meet the reinforcement requirement on the deformation of the surrounding rock masses of the fully mechanised caving coal roadway with large cross-section under the intense mining condition needs. Therefore, the newly developed high pretension cable truss controlling system is proposed to control the surrounding rock masses of the roadway. Besides the previously mentioned cable-connecting and locking instrument cable truss system, the multiple cable bolt-ring beam and the cable bolt-channel steel stretchable cable truss controlling system are also developed Meanwhile. This controlling system has the advantages that the plastic range that can be controlled is large and the anti-shearing performance is strong. Furthermore, it can response rapidly on the asymmetric character of the roadway surrounding rock mass deformation. It can also apply effective controlling on the asymmetric character of the roadway surrounding rock mass deformation.

4.4.4.1 The cable truss structure with multiple cable bolts-steel rebar composite ring beam and the asymmetric reinforcement principle

The multiple cable bolts-steel rebar composite ring beam structure is a continuous asymmetric steel rebar and multiple cable bolt composite structure that can be installed in the roadway roof. The side in which the cable bolt reinforcement density is larger is along the coal pillar. Each asymmetric steel rebar and multiple cable bolt composite structure is composed of the asymmetric steel rebar bearing beam and the multiple cable bolts which are connected with it and fixed into the deep area of the roof. Among them, the single roof cable bolt which is connected to two sides of the steel rebar bearing beam is installed inclinedly. It is fixed into the stable zone which is above the coal pillar, as shown in Figure 4-44. The asymmetric layout mechanism at this place: The cable bolt reinforcement density of the roof along the coal pillar side is larger than that along the entity coal side. Therefore, the reinforcement strengthening can be conducted on the roof along the

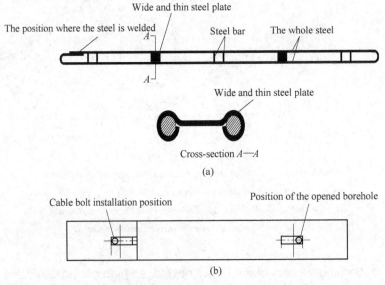

Figure 4-44 The cable truss structure with multiple cable bolts-steel rebar composite ring beam
(a) The diagram showing the structure of the steel ladder beam;
(b) The diagram showing the position of the opened hole in the channel steel (unit: mm)

weak coal pillar side. The cable bolt anchorage point is located in the rock masses that are subjected to the compressive stress along three directions and are not easy to be damaged. Therefore, it is not easy to be influenced by the bed separation and deformation of the roof that is above the coal roadway. Consequently, it provides the reliable and stable load bearing basement for developing the high anchorage force.

In the cable bolt-steel rebar composite ring beam truss system, the design of the rectangular half round cable bolt hole on the thin steel plate at two ends is beneficial for controlling the vertical sinking movement of the roof. Meanwhile, it also has the strong adaptability on the violent horizontal movement of the roadway roof. It can avoid the permanent failure problem of the structure which is resulted by the occurrence of bending when the composite structure of the cable bolt and the W-shaped steel belt or the composite structure of the cable bolt and the channel steel. The connecting component of the steel beam integrates restricting the subsidence of the roof and the function of adapting to the horizontal movement of the rock strata. It promotes the adapting ability and anti-failing ability of the cable truss system in the horizontal movement process of the rock strata.

4.4.4.2 The cable truss structure of the cable-channel steel with the extension-type beam and its basic principles

The device of the extension-type cable truss is shown in the Figure 4-45. This structure is composed of two cable bolts and the short channel steel composite beam that is closely attached on the roof. In this structure, the short channel steel composite beam is composed of the main beam, the assistant bema and 2~3 U shaped clamps. The assistant beam is coupled in the main beam and they are overlapped and coincided for a certain length. Furthermore, the U-shaped clamps are used to fix them together.

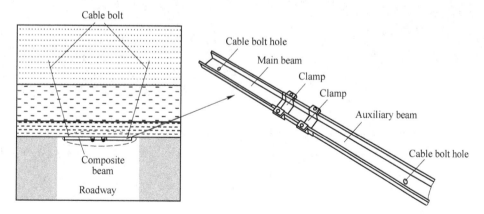

Figure 4-45 The structural diagram showing the extension-type cable truss

For the structure of the extension-type cable truss, it is the cable truss structure in which the two-section high strength short steel beam and the sunken connection component are regarded as the main part. Its structure is simple and the usage is convenient. According to the magnitude of the acting force that is generated by the horizontal movement of the roadway roof rock strata, the composite beam adjusts its self-length, arriving at the purpose of restraining or adjusting the horizontal squeezing-loose expanding deformation of the roadway roof. Furthermore, it guarantees that the composite beam does not have the phenomenon of the torsional tearing and rupture, and the disconnected failure. Overall, the cable bolt-channel extension type beam cable truss can restrict the roof sinking and the horizontal squeezing deformation. Furthermore, each component can bear the load synergistically. It can adapt to the full process of squeezing-stabilising-loose expanding which is resulted by the horizontal deformation failure of the coal mass and rock mass in the roof. It has the function of constant resistance along two directions for the horizontal deformation failure of the roof.

4.4.5 The in-situ reinforcement industrial experiment on the fully mechanised caving coal roadway with large cross-section and intense mining

Based on the above-mentioned asymmetric failure mechanism and controlling principle of the fully mechanised caving gob-side coal roadway, together with the numerical simulation calculation, mechanical analysis and engineering comparison, it is comprehensively determined that the reinforcement scheme of the ventilation roadway 20103 is shown in Figure 4-46. Meanwhile, the numerical simulation software FLAC^{3D} was used to optimise the eccentric distance, the length and the span of the cable beam truss. The results show that when the eccentric distance of the cable beam truss, the length and the span of the cable beam truss are 400mm, 8m and 1.6m respectively, the surrounding rock mass deformation and the failure depth of the plastic zone are minimal.

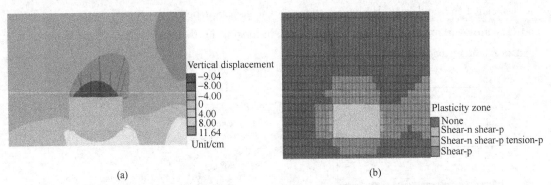

Figure 4-46 The calculating results of FLAC^{3D} when the cable beam truss span is 1.6m
(a) The vertical displacement distribution diagram of the roadway; (b) The plastic zone distribution diagram of the roadway

4.4.5.1 The reinforcement scheme design of the ventilation roadway 20103

A Roof reinforcement

a The rock bolt reinforcement
The rock bolt is the left lateral threaded high strength rock bolts without longitudinal ribs (ϕ20mm× 2500mm). The anchorage length is 950mm. The interval and the spacing are 1000mm×900mm.

In each row, there are 6 rock bolts. The intersection angle between the roof corner rock bolt along the coal side and the vertical line is 15 degrees. As for the rest roof rock bolts, they are installed perpendicular with the roof.

b The cable bolt reinforcement

For the roof cable bolt reinforcement, it is the joint reinforcement form which is composed of "Asymmetric multiple cable bolt steel beam truss" and the "High pretension cable bolt truss".

(1) The asymmetric multiple cable bolt steel beam truss.

For the asymmetric multiple cable bolt steel beam truss system, the high strength pretension steel strands ($\phi 17.8mm \times 8250mm$) are used. The spacing is 1800mm and there are three cable bolts in each row. The interval is 1500mm. Along the coal pillar side, the distance between the cable bolt and the roadway side is 800mm. Along the entity coal side, the distance between the cable bolt and the roadway side is 1800mm. The intersection angle between the cable bolt borehole that is adjacent to the roadway two sides and the vertical line of the roof is 15 degrees. As for the cable bolts in the middle, they are perpendicular with the roof.

(2) The high pretension cable truss.

The high strength pretension steel strands ($\phi 17.8mm \times 8250mm$) are used. The depth of the cable bolt borehole is 7m and the diameter of the drilled borehole is 28mm. The anchorage length is 1235mm. The spacing is 1440mm. Its bottom span is 2.1m. The distance between the cable bolt borehole and the reinforced coal side is 1.75m. The intersection angle between the cable bolt borehole and the vertical line is 15 degrees.

B The entity coal reinforcement

The normal metal rock bolts ($\phi 18mm \times 2000mm$) are used. In each row, there are 4 rock bolts installed. The rock bolt interval and spacing are 950mm and 900mm respectively. The distance between the top rock bolt and the roof is 250mm. The distance between the bottom rock bolt and the floor is 450mm. For the rock bolt that is adjacent to the roof, it inclines upwards for 15 degrees. As for the rock bolts that are adjacent to the floor, they incline downwards for 15 degrees. As for the rest rock bolts, they are perpendicular with the roadway side.

C The coal pillar reinforcement

The threaded steel rock bolts ($\phi 20mm \times 2500mm$) are used. In each row, there are 4 rock bolts. The interval and spacing of rock bolts are 950mm×900mm. The distance between the top rock bolt and the roof is 250mm. The distance between the bottom rock bolt and the floor is 450mm. For the rock bolts that are adjacent with the roof, they are inclined upwards for 15 degrees. As for the rock bolts that are adjacent with the floor, they are inclined downwards for 15 degrees. As for the rest rock bolts, they are perpendicular with the roadway side. The specific reinforcement form and parameters are shown in Figure 4-47.

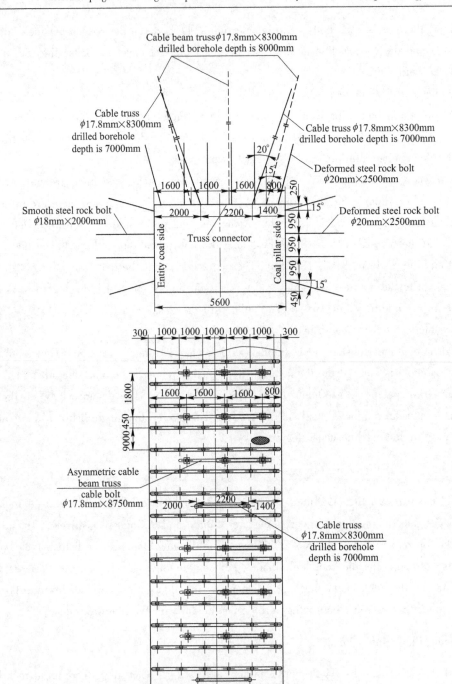

Figure 4-47 The reinforcement scheme of the haulage roadway 20103

4.4.5.2 The in-situ ground pressure observation of the ventilation roadway

The ground pressure observation results of the ventilation roadway 20103 show that during the advancing process of the working face, the deformation of the surrounding rock masses for the ventilation roadway 20103 is not large. Two sides of the roadway are relatively more sensitive to the mining influence. However, the velocity of the roadway two sides and the surrounding rock mass deformation are both in the safety range. It will have no influence on the mine production. The roof and the floor are basically not influenced by the mining activities. Overall, under the asymmetric reinforcement condition, the roadway surrounding rock mass deformation is in a pretty small state. Furthermore, the sensitivity of the roadway surrounding rock mass deformation on the mining influence is weakened. The in-situ reinforcement effect is shown in Figure 4-48.

Figure 4-48 The reinforcement effect diagram showing the ventilation roadway 20103

4.5 The gob-side roadway retaining technology and the technical optimisation and improvement in the section of the fully mechanised mining [10]

The gob-side retaining indicates that behind the coal working face, the previous mining roadway should be maintained along the edge of the gob area. A certain technical method is adopted to re-reinforce and re-support the roadway belonging to the previous section, to reserve it for the next section. As a matter of fact, it is one of the mining technologies without coal pillars. The gob-side roadway retaining has already become the main method to solve the limit overrunning of the gas in the working face where the quantity of the gas is high. If the ventilation model of the working face changes from "U" to "Y", the air volume of the working face will not only increase. Furthermore, the environmental temperature of the operating in the retreat working face will also decrease. However, there are still several challenges occurring in the gob-side roadway retaining technology under the complicated conditions such as the deep mine, the medium thick coal seam, the thick coal seam and the roadway with large cross-section. This restricts the application of the gob-side roadway retaining.

4.5.1 The surrounding rock mass controlling principle of the gob-side roadway retaining

A large number of engineering practices demonstrate that the main existing problem in the gob-side roadway retaining is that the rupture position of the main roof is uncertain. The structural state of the surrounding rock masses of the retained roadway is unclear. The bearing of the roof, wall and floor is not coordinated. The capacity of the backfill body has large deviation. Aiming at these problems, it is proposed the following surrounding rock mass controlling principles of the gob-side roadway retaining.

(1) The stress field in the surrounding rock masses around the gob-side roadway retaining can be optimised. For a certain extent, the length of the cantilever of the retained roadway roof, determines the stress and failure extent of the surrounding rock masses of the retained roadway.

(2) The strength of the fractured surrounding rock masses can be enhanced. Effective reinforcement is conducted on the fractured surrounding rock masses with low confinement. This is to improve the bearing capacity and the stability of the surrounding rock masses. The normal reinforcement methods have the reinforcement with the cable bolts and meshes, and the reinforcement with grouting.

(3) The structure of the surrounding rock masses of the retained roadway can be enhanced. The safety controlling of the roof, the reinforcement of the weaken area and the enhancing of the core load bearing area can be performed.

(4) The rock bolt bearing performance can be enhanced. The load bearing character of the rock bolt reinforcement can be enhanced synthetically. This is the development direction of the rock bolt reinforcement. Its essence is to promote the rock bolt reinforcement character curve to have the characteristics of early strength and the rapid increasing of the resistance.

4.5.2 The controlling strategy of the surrounding rock masses in the gob-side roadway retaining

4.5.2.1 The surrounding rock mass controlling technology with "three positions and one body"

The surrounding rock mass controlling technology with "three positions and one body" is constructed. It includes the back-rotating rock bolt reinforcement and cable bolt reinforcement which can resist the shearing of the roof (P_1), the auxiliary strengthened reinforcement in the roadway (P_2) and the backfilled wall body along the roadway side (P_3), as shown in Figure 4-49.

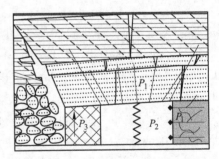

Figure 4-49 The schematic diagram showing the surrounding rock mass controlling technology with "three positions and one body"

4.5.2.2 New surrounding rock mass controlling technology with "three high" character

Upgrading of the rock bolt technology: (1) The normal smooth steel rock bolts; (2) The high strength threaded steel resin rock bolts (one high); (3) The high performance pretensioned rock bolts (two high); (4) High system stiffness and high-performance rock bolts with high strength (three high). The "three high" anchorage technology regards the high strength rock bolts and the reinforcement accessories with high stiffness as the basement. It is the reinforcement technology which regards the high pretension force as the core (high strength, high stiffness and high pretension force). This is shown in Figure 4-50.

4.5.2.3 The roof and floor stability controlling technology of the load transfer along two directions

The roof and floor stability controlling technology of the load transfer along two directions is proposed. When the deformation of the roof and the floor induces the load on the single hydraulic prop, the single hydraulic prop realises the restraint on the deformation through releasing the load and increasing the resistance based on the mobile column. This can make sure that the deformation of the roof and the floor ($\Delta_1 + \Delta_2$) can be fully controlled within the contraction range of the single hydraulic prop, as shown in Figure 4-51.

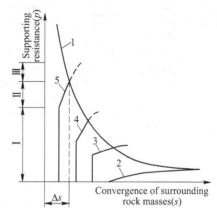

Figure 4-50 The schematic diagram showing the rock bolt operating condition and the surrounding rock mass deformation relationship
1—Typically supported surrounding rock mass relationship curve;
2—Traditional reinforced character curve;
3—High strength rock bolt reinforced character curve;
4—High performance rock bolt reinforced character curve;
5—"Three high" rock bolt reinforced character curve

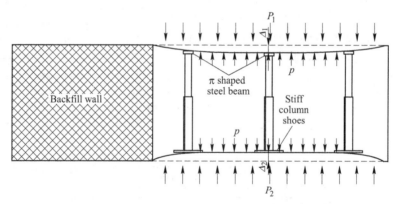

Figure 4-51 The roof and the floor stability controlling technology of the load transfer along two directions

4.5.2.4 The deep borehole pre-splitting and blasting controlling technology of the roof

The deep borehole pre-splitting and blasting controlling technology of the roof indicates that during the recovery mining process of the working face, the blasting in the deep borehole is conducted to pre-split the roof, realising the purpose of releasing the pressure in the surrounding rock masses of the retained roadway. It is applicable for the conditions that the thick and strong roof is directly covered or the thickness of the immediate roof is not sufficient, leading to the consequence that the "prescribed deformation" of the main roof is too large. The arrangement form and parameters of the blasting borehole should be determined based on the mining condition, as shown in Figure 4-52.

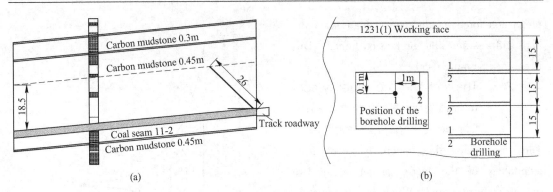

Figure 4-52   The schematic diagram showing the pre-splitting and blasting in the deep borehole

(a) The lateral view; (b) The vertical view

4.5.2.5   The integrity controlling technology of the roof that is above the backfilling area

Due to the influence of the advanced abutment pressure and the repeated support that the backfilling supports apply on the roof, the roof that is above the area that is to be backfilled may fail, loose and collapse. The controlling of the backfill wall roof and floor along the roadway is the core in the success of the gob-side roadway retaining. The commonly used integrity controlling technology of the roof that is above the backfilling area include opening the gap in advance and the reinforcement of the roof that is above the backfilling area, as shown in Figure 4-53 and Figure 4-54.

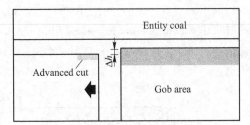

Figure 4-53   Opening the gap in advance

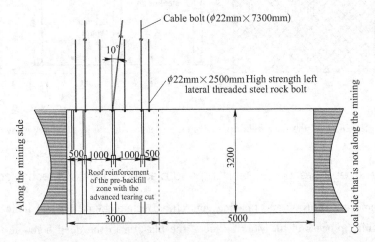

Figure 4-54   Reinforcement of the roof that is above the backfilling area

4.5.2.6   Strengthening the structure of the backfill body roof and floor

After the roadway is retained, within the range that is after the working face for 30-80m, influenced by the violent dynamic pressure, rupture and the back-rotating subsidence occur in the overlying strata. This leads to the inclination of the roof. To prevent the rotation of the backfill body in the gob-side roadway retaining, inclination of the roof and the roof cutting, the roof and the floor must be strengthened, as shown in Figure 4-55.

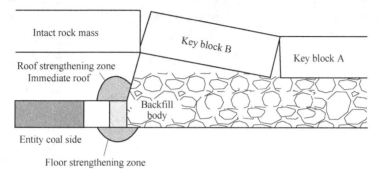

Figure 4-55 The optimisation position of the strength of the backfill body roof and floor

On the condition that after the activities of the overlying rock strata, the twice reinforcement can be conducted on the entity coal side. Itcan effectively improve the strength of the surrounding rock masses of the retained roadway. And then, it can be guaranteed that after the next working face is mined, the retained roadway can be used normally, as shown in Figure 4-56.

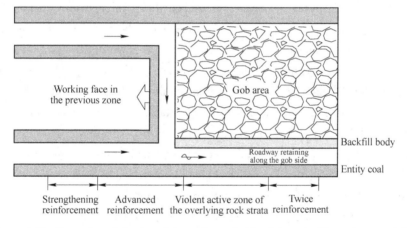

Figure 4-56 Strength optimisation of the entity coal side of the gob-side roadway retaining

4.5.2.7 The "step" controlling technology in the retained roadway

In the process of the expanding and repeated using of the gob-side roadway, the step is formed between the original roof and the newly exposed roof. Furthermore, this step has the tendency to expand. If this step is not processed in time, the roof "step" will usually become severe. Bed separation of the roof may occur and even the roof collapsing may occur. This seriously influences the safe production of the next recovery working face, as shown in Figure 4-57. The reasons to induce the "steps" include the following aspects. The service time of the retained roadway is too long. Moreover, the reinforcement strength of the roof is not sufficient. Additionally, the tunnelling is conducted along the roof of the coal seam. The expanding and tunnelling are conducted from the stopping line. Finally, the reinforcement of the original roof is failed.

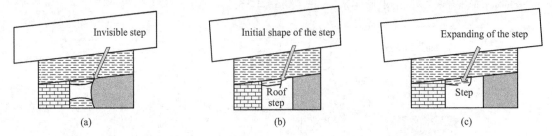

Figure 4-57 The forming process of the retained roadway roof "step"
(a) After the roadway is retained; (b) The initial period of the expanding of the roadway retaining;
(c) The latter period of the expanding of the roadway retaining

4.5.3 The technique of the gob-side roadway retaining system

The appropriate technique of the gob-side roadway retaining can not only guarantee the safety of the retained roadway, but also reach the effect of improving the efficiency. The specific technique includes the following aspects:

(1) Strengthening the roadway before the mining activities. Through strengthening the roof and the entity coal side, the roadway reinforcement strength can be improved and the safety of the retained roadway can be guaranteed.

(2) Pre-splitting and blasting in the deep borehole. The pre-splitting and blasting are conducted in the deep borehole in the roof in front of the working face. This can decrease the length of the roof cantilever for the retained roadway. Then, it decreases the violent extent of the activities of the overlying rock strata for the retained roadway.

(3) Opening the gap in advance. It can improve the strength of the roof which is above the area that is to be backfilled. Then, the roof collapsing can be prevented. Furthermore, for the roadway retaining, the backfilling along the gob-side can be conducted successfully.

(4) Backfilling along the roadway side and constructing the backfilling wall.

(5) The reinforcement is strengthened for the retained roadway. In the roadway, the entity coal side is strengthened. In the latter period of the roadway retaining, the immediate reinforcement is conducted. This is to release the influence of the violent activities of the overlying rock strata.

The backfilling technique along the roadway side, mainly includes the following contents:

(1) Dimension of the backfilling wall.

The influencing parameters of the backfilling wall include the strength in mining, the buried depth, the dip angle of the coal seam, the mining height and the properties of the rocks in the roof and the floor.

(2) Position of the backfill wall.

The position of the backfill wall includes the gob-side roadway retaining at the original position, the gob-side roadway retaining at the semi-original position, the roadway retaining in the roadway, the roadway retaining along the side. The specific advantages, disadvantages and the application condition are shown in the Table 4-1.

Table 4-1 The position of the different backfill wall and the advantages and disadvantages

Backfill position	Advantages	Disadvantages	Application condition
Gob-side roadway retaining at the original position	The width of the retained roadway is large	The cross-section of the roadway is large and it is difficult to maintain	There is no false roof. The rock properties of the roof is favourable and it is integrated
Gob-side roadway retaining at the semi-original position	The width of the retained roadway is relatively large	The appearance of the ground pressure is relatively severe and the cross-section of the roadway is small	The strength of mining is high and the roadway is retained for a long time
Roadway retaining in the roadway	The width of the retained roadway is small and it is easy to maintain. The top roof side is stable	If the roadway deforms, it is difficult to fulfil usage	The rock properties of the roof are favourable and it is integrated. The ground pressure is small and it is not necessary to solve the gas problems
Roadway retaining along the side	The backfill body is easy to be constructed. The other reinforcement methods are not necessary to be implemented. The backfill body is easy to maintain stability	For the recovery mining of the adjacent working face, the roadways must be re-tunnelled	The volume of the gas burst is small

(3) The backfill equipment along the roadway side.

The backfill along the roadway side includes the hydraulic backfill support, the concrete backfill pump and the pipe lines. The plane graph of the backfill system along the gob side is shown in Figure 4-58.

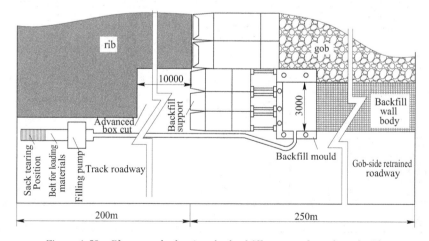

Figure 4-58 Plane graph showing the backfill system along the gob side

(4) Backfill procedures.

For the backfill technique along the roadway side, the specific procedures are shown in Figure 4-59. This technical procedure has the following characters. The operation is simple. It is reliable and continuous. Furthermore, the backfilling velocity is rapid.

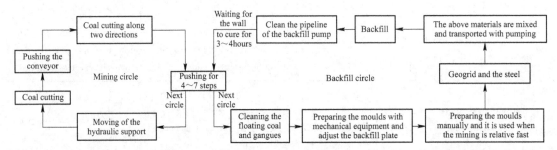

Figure 4-59 The technical procedures of the backfilling along the roadway side

4.5.4 In-situ tests of the gob-side roadway retaining

The Panyi Coal Mine in Huainan mainly exploits the coal seam 11-2. The average buried depth is larger than 800m. In this area, the maximum horizontal main stress is generally distributed along the east-west direction, which is around 33.9~36.4MPa. The vertical stress is generally around 12.1~15.4MPa. For the working face 1231(1), the dip direction is along the east-west direction. The length along the strike direction is 1367m and the width is 240m. The mean coal seam thickness is 2.45m and the mean dip angle is around 4 degrees. The relative gas emission quantity is 14m^3/t and the water inflow is small. The roadway retaining is conducted on the railway roadway of this working face. The specific arrange is shown in Figure 4-60.

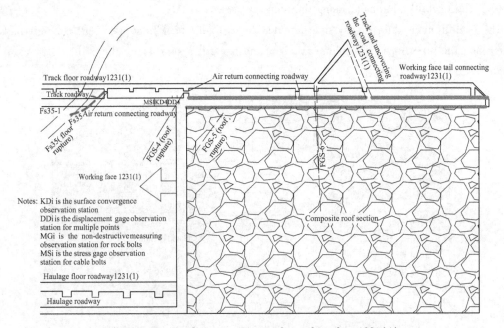

Figure 4-60 Roadway remaining in the working face 1231(1)

4.5.4.1　Design of the reinforcement scheme of the railway roadway

Considering the practical geological production condition of the Panyi Coal Mine in Huainan, the numerical simulation calculation, mechanical analysis and the engineering comparison are comprehensively used to determine the following reinforcement scheme of the railway roadway of the working face 1231(1).

The reinforcement method of the roof for the railway roadway: the cable bolt reinforcement and the mesh support are used. 6 left ribbed threaded steel rock bolts without longitudinal ribs ($\phi22mm\times2500mm$) are used. The interval and spacing are 900mm×800mm. The cable bolt with the parameter of $\phi22mm\times7300mm$ is used. Cable bolts are installed with the 3-3 form. The interval and the spacing are 1100mm×800mm. Along the strike direction, the box iron 14# ($L=2800mm$) matched with the cable bolts ($\phi22mm\times7300mm$) are used. Reinforcement of the roof before mining: the hollow grouted cable bolts ($\phi22mm\times7300mm$) are installed additionally → grout spraying → grouting (super fine cement). In front of the working face for 100m, construction is conducted, as shown in Figure 4-61.

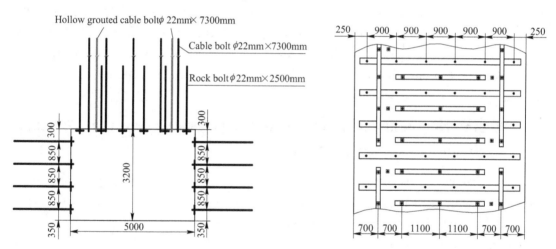

Figure 4-61　The cable bolt and meshing support technical scheme for the railway roadway roof
(Notes: the black colour indicates the original support and the red colour indicates the strengthened support before mining)

Reinforcement of the railway roadway two sides: The rock bolt reinforcement and the meshing support are used. 4 rock bolts ($\phi22mm\times2500mm$) are set. The interval and spacing are 850mm×800mm. For the entity coal side, the reinforcement is conducted before mining. The cable bolt and the beam support method is used. Along the strike direction, there are three cable bolt beams. The cable bolts ($\phi22mm\times7300mm$) are used. The interval and the spacing are 1400mm×1000mm. Meanwhile, in the retained roadway, different methods are used to strengthen the support, such as the single column, the I-beam, and the timber crib matched with the timber support. This is shown in Figure 4-62.

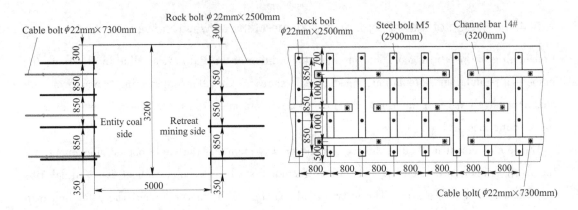

Figure 4-62 The reinforcement scheme of the railway roadway two sides

4.5.4.2 Analysis of the ground pressure observation results

The in-situ ground pressure observation data show that for the advancing working face, the convergence along the retreat mining side is larger than that of the entity coal side. The convergence along the retreat mining side is the main part. However, in the retained roadway, the convergence along the entity coal side is the main part. In the advancing working face and the retained roadway, the floor convergence is both larger than the roof convergence. And the floor deformation is the main part. Overall, the roadway roof sinking and the roadway two side deformation are both in the safety range. This fulfils the using requirement of the in-situ production.

4.6 Engineering design and practice of the pressure relieving in the high stress and weak coal roadway

In Chinese coal mines, the coal body strength of a great number of thick coal seams is low and the bottom strata are soft, and the fully-mechanized longwall top-coal caving mining has developed rapidly and been used widely. Connected to a caving mining face, the panel entries are generally laid out in a coal seam and along its floor. Apparently, the roof and both sides of the panel entries are all coal, and the majority of these panel entries are soft rock roadways. Furthermore, due to coal mining, the tail entries along the gob area are dually affected by the immobile abutment pressure distributed along the coal seam dip and the moving one distributed along the coal seam strike and in front of coal faces. Therefore, the cross section of this kind of roadways quickly lessens and the roadway supports are damaged on a large scale, which has a very bad influence on the fully-mechanized longwall caving mining with high output, efficiency and safety.

4.6.1 Geological and productional conditions

The average dip angle of coal seam No. 13 in Xinji coal colliery is 12°. Coal panel No. 1309 is located at $-345 \sim -442$m, and the corresponding soil surface is at about +26m above sea level. The geological log of coal seam No. 13 and its roof and bottom is shown in Figure 4-63.

Serial number	Stratum name		Stratum thickness /m
1	Arenaceous argillite		9.70
2	Argillite		3.64
3	Siltstone		2.59
4	Medium-grained sandstone		3.29
5	Arenaceous argillite		2.20
6	Coal seam No.13		8.90
7	Arenaceous argillite		3.02
8	Arenaceous argillite		9.70

Figure 4-63 The geological log in panel No. 1309

The following formula is the criterion for a stratum to be soft in a coal mine.

$$\eta = \gamma H/R_c \geqslant 0.4 \sim 0.5 \tag{4-7}$$

Where η——stratum stability coefficient;

γ——average density of the overlying strata, kN/m^3;

H——depth, m;

R_c——uniaxial compressive strength of the stratum rock, kN/m^2.

The η of coal seam No. 13 is 1.35~1.47, so this coal seam is very soft.

The destressing engineering was first applied to the tailentry along gob area and at fully-mechanized longwall top-coal caving mining panel No. 1309 in Xinji coal mine. This tailentry is located in coal seam No. 13 and along its floor, and belongs to a soft rock roadway influenced by high abutment pressure. The measurement results of the tailentry deformation beyond the destressing zone indicated that the maximum roof-to-floor convergence reached 2347mm and the maximum convergence between two coal sides amounted to 1962mm. Although the tailentry was repaired or rebuilt 3~5 times and a large amount of manpower, material and finance resources were wasted, the tailentry cross section might deform and decrease from the original 9.25m^2 right after excavation to 0.5~2.0m^2 or even less value. As shown in Figure 4-64, the support——surrounding rock system was severely damaged, which resulted in high ventilation resistance, high velocity of air-flow, much coal dust in the air-flow and a very narrow and low sidewalk in the tailentry. The equipment relocation in the tailentry and upper face end maintenance were extremely difficult.

4.6.2 Simulation and design of the destressing scheme

Since the coal seam in caving mining face No. 1309 is 8.90m thick, the new destressing project of the tailentry along gob area can be designed as shown in Figure 4-65. The destressing entry and tailentry in coal body are arranged along the roof and floor of the coal seam respectively. The destressing entry is located in the coal body above the tailentry, and the loosening blasting zones

Figure 4-64 The tailentry damage

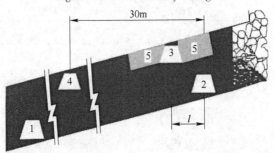

Figure 4-65 The destressing project of the tailentry in panel No. 1309

1—Headentry; 2—Tailentry along gob area; 3—Destressing entry; 4—Gas draining-out roadway;
5—Loosening blasting area; l—The distance between two entries

are within both coal sides of the destressing entry, and then the tailentry's surrounding rocks are not artificially broken due to the loosening blasting and can be destressed and situated in the low stress area. Consequently, the situation of serious deformation and failure, together with extremely difficult maintenance of the tailentry's surrounding rocks, will substantially be changed. The trapezoidal cross section dimensions of the destressing entry or the tailentry along gob area are upper width × lower width × height = 3.2m × 4.2m × 2.5m. Roadways are supported with trapezoidal supports of mine I-steel No. 12, and the set spacing is 0.5m.

During the numerical simulating calculation process, the stress state and displacement field and deformation rate of the tailentry's surrounding rocks for every destressing scheme are monitored and recorded. The stress and deformation state of the tailentry's surrounding rocks without destressing engineering is shown in Figure 4-66, and the tailentry's surrounding rocks are located in the high stress area and it's total deformation is very serious. Nevertheless, the numerical simulating calculation results indicate that the destressing engineering greatly decreases the stress transmission from overlying strata down to surrounding rocks of the tailentry along gob area and makes the high stress shift from the tailentry's surrounding rocks to the deeper coal body along the coal seam dip, which benefits the control of the plastic deformation and volume expansion in the tailentry's surrounding rocks.

Based on the analysis of numerical simulation calculation results and destressing effects of different

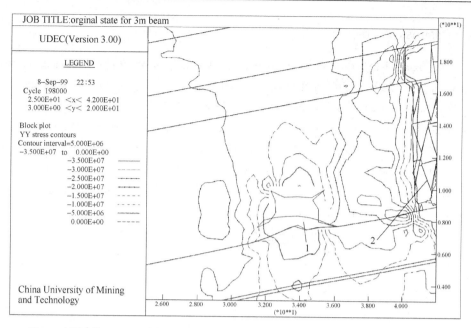

Figure 4-66 The stress of the tailentry's surrounding rocks without destressing engineering

1—Tailentry along gob area; 2—Gob area

destressing schemes, the factual geological and productional conditions and the safe production request in Xinji coal mine, the optimum design of the destressing engineering scheme is that $l = 6.0$m, $d = 5.0$m, and the extent of loosening blasting should be high. The stress and deformation state of the tailentry's surrounding rocks in the designed scheme is shown in Figure 4-67.

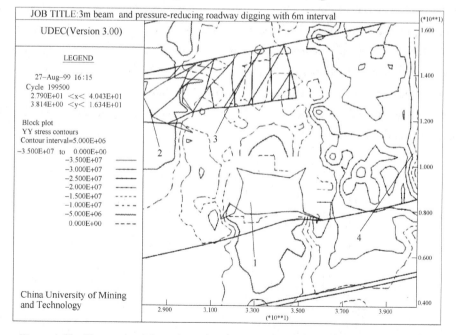

Figure 4-67 The stress of the tailentry's surrounding rocks with destressing engineering

1—Tailentry along gob area; 2—Destressing entry; 3—Loosening blasting area; 4—Gob area

4.6.3 Destressing practice

As shown in Figure 4-68, the field operation of the tailentry's destressing engineering at caving mining panel No. 1309 in Xinji coal mine was performed according to the following steps.

(1) The connecting roadway in coal seam No. 1309 and along its roof was first driven from the gas draining-out roadway to the designed location of the destressing entry. Then the destressing entry and exploration roadways were driven. The distance between the adjacent exploration roadways is 10m, and the exploration roadways are 5~6m long. The exploration roadways can improve the loosening blasting effect, and ensure that the loosening blasting zone has no connection with the upper neighboring gob area.

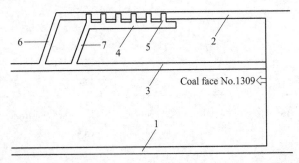

Figure 4-68   Layout of the roadway system in panel No. 1309
1—Headentry; 2—Tailentry along gob area; 3—Gas draining-out roadway; 4—Destressing entry;
5—Exploration roadway; 6—Connecting roadway to the tailentry; 7—Connecting roadway to the destressing entry

(2) Subsequently, in order to simplify the roadway system management, control the gas emission and prevent spontaneous combustion of coal body in the destressing entry, the mining engineers and workers ought to carry out the loosening blasting of coal body in both destressing entry sides and withdraw the trapezoidal I-steel supports. The blast hole pattern on each coal side of the destressing entry is shown in Figure 4-69. The blast hole depth and diameter in each coal side are 5.0m and 36~42mm respectively. In each blast hole, the powder charge is 2450g and the stemming material length is 1.5m.

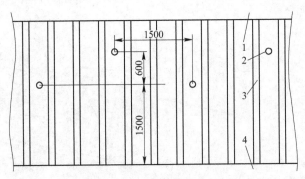

Figure 4-69   The blast hole pattern on the coal side
1—Entry roof; 2—Blast hole; 3—Splayed leg; 4—Entry bottom

(3) After the above engineering was finished, the destressing entry must be sealed in time, and the tailentry along gob area was finally excavated.

The roof-to-floor convergence and the convergence between two coal sides were measured in the tailentry along gob area and connected to caving face No. 1309. The measurement results showed that after the destressing technology was adopted, the deformation and failure degree of the tailentry's surrounding rocks noticeably lessened, and the damage rate of tailentry supports lowered by 78%. The surrounding rock stress of a tailentry along gob area can be largely decreased by use of the destressing roadway and loosening blasting in the coal body above the tailentry.

5 The challenging problems of deep mining and dynamic pressure study

5.1 The challenging problem of surrounding rock mass controlling in the deep mining and the engineering technology [11]

The coal exploiting in the deep area is the challenging problemthat is confronted by multiple coal production countries in the world. The mining depth of German, Russia and Portland is all beyond 1000m. For German, it is 1750m. In China, the mining depth of 47mine sites is more than 1000m. The deepest one is 1501m. The particular environment of the deep rock masses makes the rock mass character and the ground pressure appearance character have the essential variation. Due to the speciality of the mechanical character of the deep rock masses, and the complexity of the reserve environment, the following difficulties in the roadway surrounding rock mass controlling are resulted. The mine stress is high and the structural stress is complicated. The mining influence is violent. The deformation of the roadway surrounding rock masses is large. The floor heave is violent. The roof collapsing and the rib spalling occur. The damaging of the supporting body is serious. The roadway deformation has the burst tendency.

5.1.1 The reinforcement technology of the roadways in the deep area in China and overseas countries

5.1.1.1 The reinforcement methods of the roadway in the deep area in overseas countries

In German, the cross-section of the roadway in coal mines usually has the arch shape. The cross-section of the roadway is large and the cross-section area is usually approximately 30m^2. When it comes the support of the roadway, the U-shaped steel retractable supports account for 65%. And the combined support with the supports and rock bolts account for 30%. The coal mines in German particularly emphasises the importance of the backfill behind the supports. As a matter of fact, the backfill behind the supports can make sure that the supports can fully contact with the surrounding rock masses, which is beneficial for improving the stiffness of the load transfer medium. Then it can be guaranteed that the resistance of the arch-shaped supports can be transferred to the roadway. After the backfill is conducted behind the supports, the load bearing capacity of supports can be increased more than twice.

The combination of the supports and rock bolts has two different types, as shown in Figure 5-1. The first type means that after the excavation, the rock bolts are installed. At the position that is 10~50m far away from the tunnelling working face, the supports are installed. Then, the backfill is conducted after the supports. The second type means that after the exaction, the supports are installed and then the backfill is implemented. At the position that is 20~100m far away from the tunnelling working face, the rock bolts are installed.

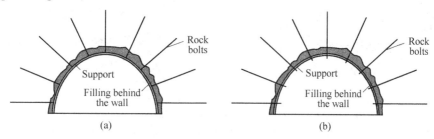

Figure 5-1 The combined support with supports and rock bolts
(a) The rock bolts and the U-shaped steel supports; (b) The U-shaped steel supports and the rock bolts

5.1.1.2 The reinforcement technology in the deep roadways in China

A The development process of the roadway reinforcement technology

In China, the roadway reinforcement technology mainly experiences the following development process, namely the casting support, the steel support, the low strength rock bolt reinforcement, the high strength rock bolt support and the high pretension and high strength rock bolt reinforcement, as shown in Figure 5-2.

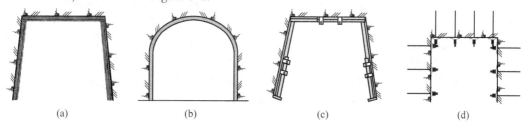

Figure 5-2 Development of the deep roadway support and reinforcement
(a) Timber support; (b) Casting support; (c) Steel support; (d) Rock bolt reinforcement

B The reinforcement forms of the deep roadways

The reinforcement forms of the deep roadways in China mainly include the following: (1) High strength rock bolt and cable bolt reinforcement; (2) The U-steel retractable support and the steel pipe concrete support, as shown in Figure 5-3; (3) grouting reinforcement (such as the cement and chemical materials); (4) pressure relieving method (mining along the upward direction, cutting the gap, borehole drilling, blasting and hydraulic fracturing); (5) combined support and strengthening (rock bolts + supports; rock bolts + supports + grouting).

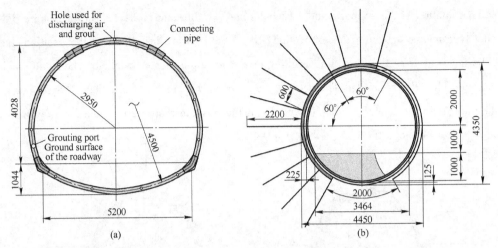

Figure 5-3　The reinforcement forms of the deep roadways

(a) Steel pipe and concrete support；(b) U-steel retractable support and rock bolts

5.1.2　The development of the deep roadway reinforcement theory

5.1.2.1　Traditional deep roadway reinforcement theory

For the traditional deep roadway support theory, it is based on the roadway support surrounding rock mass response curve which is set on the excavation surface, as shown in Figure 5-4.

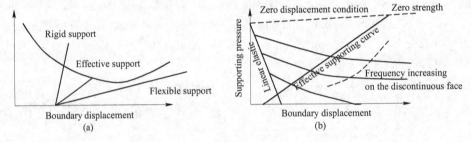

Figure 5-4　The traditional roadway support surrounding rock mass response curve

(a) Continuous rock masses；(b) Discontinuous rock masses

Combined support: It is extremely difficult to effectively control the surrounding rock mass deformation of the deep roadway with only a single support method. Therefore, two kinds of methods or more than two kinds of methods should be adopted toperform the combined support. Secondary support: The single support is quite difficult to control the large deformation of the deep roadway surrounding rock masses. Therefore, the secondary support should be used. First, the flexible support should be used and then the rigid support should be used. The pressure relieving and the resistance should be conducted, as shown in Figure 5-5.

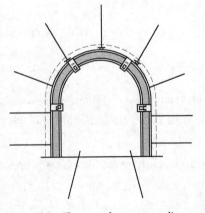

Figure 5-5　The secondary support diagram

For the deep roadways, after the secondary support, there are still failures. Then, the third support, the fourth support and even more supports are needed.

5.1.2.2 New understanding on the reinforcement effect of rock bolts on the deep roadways

A New understanding on the rock bolt reinforcement

(1) The rock bolts include the parts that are inserted into the surrounding rock masses (rock bolt bar and the anchorage grout) and the surface components (the bearing plate, the steel belt and the mesh).

(2) The rock bolt reinforcement mechanism has significant difference with the support. Therefore, the hydraulic support-surrounding rock mass response curve cannot be used again. The suffered load of it on the surrounding rock masses, is shown in Figure 5-6.

(3) The main effect of rock bolts: They can control the discontinuous and inharmonious volume expanding deformation. The pretension plays a core effect. The high pretension can expand effectively. It can restrain the volume expanding deformation.

(4) After the roadway is excavated, immediate support should be conducted. Enough pretension should be applied and then the effect of the rock bolt reinforcement is best. After the surrounding rock masses generate a certain deformation, the reinforcement effect will apparently be influenced if the rock bolt reinforcement is conducted.

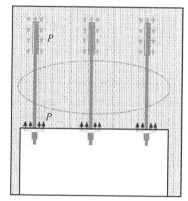

Figure 5-6 The suffered load of the surrounding rock masses in the rock bolt reinforcement

(5) The secondary reinforcement theory is not applicable for the rock bolt reinforcement.

B The reinforcement stress field

The reinforcement stress field indicates the stress field that is generated by the rock bolt reinforcement in the surrounding rock masses. The distribution diagram is shown in Figure 5-7. The in-situ stress field, the mining induced stress field and the reinforcement stress field constitute the combined stress field. The mutual acting and coordinating of those "three fields" are the core of the mining induced roadway reinforcement.

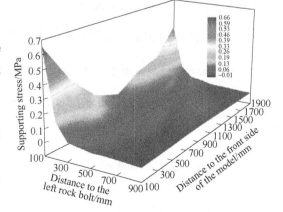

Figure 5-7 The reinforcement stress field distribution diagram of a single rock bolt

C The practical suffered load state of the underground rock bolts

(1) It is not perpendicular. And it is not the realistic tension;

(2) It is subjected to tension, bending, torsion and shearing;

$$\sigma_i = \sqrt{\left(\frac{4P_t}{\pi d^2} + \frac{32M}{\pi d^3}\right)^2 + 768\left(\frac{M_t}{\pi d^3}\right)^2} \tag{5-1}$$

(3) The suffered load of the rock bolt bar is extremely not uniform. Most of the rock bolts are working under the yielding state;

(4) High strength, high elongation and high bump toughness, as shown in Figure 5-8.

5.1.2.3 The high pretension reinforcement theory with one time in the deep roadways

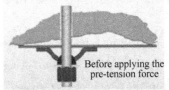
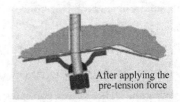

Figure 5-8 The practically suffered load state of the underground rock bolts

Rock bolts can control the discontinuous and discordant deformation of the anchorage area. They can maintain the integrity of the surrounding rock masses and weaken the decreasing of the strength. The rock bolt pretension and its effective expanding are playing the crucial effect. The rock bolt reinforcement system has enough elongation rate and bump toughness. On one aspect, it makes the continuous deformation releasing of the surrounding rock masses. On the other aspect, it avoids the local failure. If the surrounding rock masses are fractured, it is not beneficial for the rock bolt pretension. When the working resistance is expanded, grouting should be conducted. In the deep of the roadways, the high pretension and the high strength rock bolt should be used to conduct the reinforcement. This is to realise the one-time reinforcement. Rock bolts cannot effectively control the overall deformation of surrounding rock masses. Therefore, they should be combined with metal supports to form the combined support.

5.1.3 The deep roadway reinforcement technology and typical example

5.1.3.1 The deep roadway reinforcement technology

(1) Based on the geological mechanical measuring, the set of coal roadway reinforcement technologies, which regard the anchoring and grouting reinforcement as the core, is shown in Figure 5-9.

(2) Based on the geological mechanical measuring, the roadway reinforcement design is shown in Figure 5-10.

5.1.3.2 The typical example of the deep roadway reinforcement

Xinwen is the deepest mine site in China. The average mining depth is more than 1000m and the

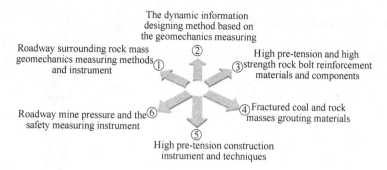

Figure 5-9　The set of coal roadway reinforcement technology

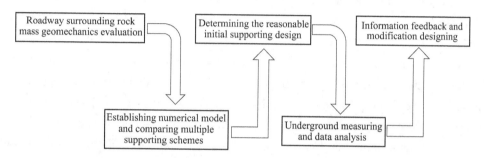

Figure 5-10　The roadway reinforcement process-dynamic information design

deepest part is 1501m. The geological structures are complicated. The in-situ stress is high. The dynamic accidents occur frequently. The roadway deformation is large and the creeping is obvious. Furthermore, the deformation is all-dimensional and asymmetric. The floor heave is more than 1m, accounting for 80%.

A The geology and production condition in the experimental site

In the Huafeng mine, the buried depth of the east rock roadway (-1180) is 1274m. The roadway crosses the siltstone. The immediate roof is the medium-size sandstone. The beddings are developed and fractured. For the maximum horizontal main stress, the minimum horizontal main stress and vertical stress, they are 29.3MPa, 14.7MPa and 31.9MPa respectively. It has the arch cross-section. The width is 5.2m and the height is 4.5m. The deformation of the original reinforcement roadway is large and the floor heave is violent.

B Grouted cable bolts

The grouted cable bolts arethe developed cable bolting and grouting technology for the fractured surrounding rock masses. They are composed of the steel strands, the locking instrument, the arch shaped bearing plate with holes, the grout blocking plug. The resin cartridge is used to anchor the end of the cable bolt. Then, the pretension is applied. After that, the cement or the chemical grout are used to fully anchor. The grouting parameters are controlled. For the fractured surrounding rock masses, the grouting is used to reinforce. Its structure is shown in Figure 5-11.

C The roadway reinforcement scheme

(1) The reinforcement method. The rock bolts and the grouted cable bolts are used together to perform the combined reinforcement. The high strength rock bolts (B600 and 22mm) with a length of 2.4m are used. Also, the steel bearing plate and the steel mesh are used. For the roof and side cable bolts, the high strength cable bolts (22mm and 5.3m) are used. The resin grout is used for the point anchor. Then, the cement is used to conduct the grouting. As for the floor, they are anchored with cable bolts. The diameter of the drilled borehole is 56mm. The "bottom anchorage and top grouting" is used, as shown in Figure 5-12(a).

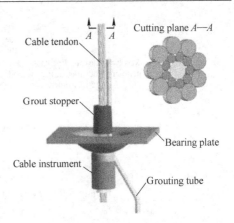

Figure 5-11 The structure of the grouted rock bolt

(2) The strengthening method. First, the cable bolt is used to mix the resin cartridge to conduct the anchoring. Then, the pretension is applied. Due to the fact that the diameter of the drilled borehole in the floor is relatively large, the special mixing end is set on the cable bolt. Then, the gangues are used to backfill. The concentrated grouting is conducted for the rock bolt free section, as shown in Figure 5-12(b).

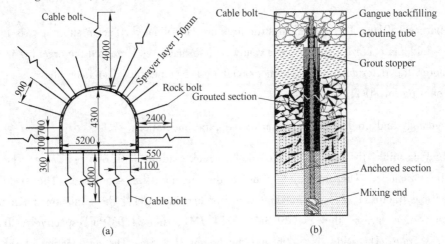

Figure 5-12 The rock roadway (-1180) reinforcement scheme in the Huafeng Mine

(a) The rock roadway (-1180) reinforcement position in the Huafeng Mine;
(b) The pretension cable bolt reinforcement in the floor

D The roadway reinforcement effect

The convergence of the roof and floor, and the two sides, are 76mm and 97mm. The roof sinking is 8m and the floor heave is 68mm. After the roadway surface deforms for 60 days, it becomes stable. The high strength pretension rock bolts on the whole cross-section, and the grouted cable bolts effectively control the large deformation and violent floor heave of the roadways with a thousand metres in depth, as shown in Figure 5-13.

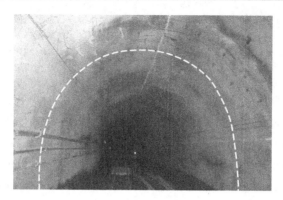

Figure 5-13   The rock roadway (-1180) reinforcement effect diagram in the Huafeng Mine

5.2 The challenging problem and the solving practices of the floor heave in the weak rock roadways in the deep mine [12~16]

The roadways are influenced by the tunnelling and retreat mining. The roof, floor and two sides will generate deformation and the displacement towards the roadway will occur. Also, the floor will heave. This phenomenon is called as the floor heave. The Lianghuai mine sites are the area that the problems of the coal mine roadway floor heave frequently occur. In the Lianghuai mine sites, the geological conditions are complicated. Also, there are many fault structures. In the deep area, the rock strata fractures are developed. The weak rocks are distributed widely. For the Lianghuai mine sites, the rock strata that belong to the weak rock type have a larger percentage in the coal seam rock strata. The tunnelling and mining disturbing is violent. With the rapid improving of the mine production centralisation and the mechanisation, the production scale expands. This leads to the consequence that the mining influence of the working face is violent. The roadway arrangement system of the high-gas mines is complicated. The tunnelling of the roadway groups disturbs mutually. The influence is violent. In recent ten years, the mining depth of the newly constructed mines continuously expands in the Huaibei Mine Site. The in-situ stress increases gradually. Its roadway floor heave is shown in Figure 5-14.

Figure 5-14   The floor heave of the roadway in the Huaibei Taoyuan Mine

With the exploiting in the mine sites in China continuously develops into the deep area, the roadway surrounding rock masses show the characters, such as weakness, fracture, loosing, expanding, poor bonding and high in-situ stress. Additionally, the water in the surrounding rock mass fractures are developed. The floor is also eroded by the water in the fractures. For the soft and weak floor, the degradation in water with a large range occurs. This intensifies the deformation and failure of the roadway floor. All these parameters become the main reasons leading to the floor heave in the deep coal masses and rock masses in mine sites.

5.2.1 Mechanism of the roadway floor heave

5.2.1.1 The types of the roadway floor heave

According to the type that the rock masses in the floor of the roadway are uplifted to the roadway and its failure mechanism, the floor heave can be divided into the following five modes, namely the floor heave with squeezing and flowing, the floor heave with bending and folding, the floor heave with shearing and relatively movement, the floor heave with expansion after encountering with water and the composite floor heave.

A The floor heave with squeezing and flowing

The immediate floor of the roadway is fractured rock masses. The structure of the roof and the two sides of the roadway is intact. Or the roadway is fully involved in the relatively weak rock strata. For the two sides of the roadway and the roof, effective reinforcement is adopted. Under the condition of the high stress, the effect of the compressing and stress (vertical stress and horizontal stress) of the rock masses in the two sides of the roadway makes the relatively weak rock masses be squeezed and flowed to the roadway.

B The floor heave with bending and folding

The mechanism of the floor heave with bending and folding is that the rock masses in the floor show bending and folding towards the empty area of the floor under the compressive pressure along the direction that is parallel with the joints. Then, instability occurs. On the condition that the rock masses in the floor of the roadway are bedded, the floor heave will still occur when the stress state meets a certain condition, although the floor is composed of medium strong rock masses.

C The floor heave with shearing and relatively movement

On the condition that the immediate floor of the roadway is intact and the thickness of the rock strata is thick, the rock masses in the floor are easy to show the shear failure under the effect of the high stress. Then, relative movement occurs in the formed wedge-shaped rock mass under the squeezing of the horizontal stress. As a consequence, the floor heave occurs. Due to the fact that the extent of the stress concentration in the corner of the roadway is high, the shear failure zone initially occurs in this position. With the development of the failure, it is connected with the shear failure zone. And then, the wedge-shaped failure is formed.

D The floor heave with expansion after encountering with water

The floor heave with expansion after encountering with water means that after the expansion rock masses in the floor encounter with water, the physical reaction and the chemical reaction occur. Furthermore, with the time increasing, the volume of the rock masses also increases. As a consequence, the floor heave occurs. The expansion rock masses are mainly composed of clay stones. The mineral components include the montmorillonites of which the physical characters and chemical characters are active.

E The composite floor heaves

The rock masses in the floor are influenced and restricted by many different parameters, such as the horizontal tectonic stress, the shear stress and the expansion stress. They act mutually, forming the compositefloor heave.

5.2.1.2 The principal reasons in influencing the floor heave

A The characters of the surrounding rock masses

The characters of the surrounding rock masses and the state of the structure are playing the crucial effect in determining the floor heave. These are mainly reflected in the following aspects:

(1) The structural state of the rock masses in the floor (the structure of the fracturing, the thin layer shaped structure and the thick layer shaped structure) determines the floor heave type of the roadways which is mentioned above;

(2) The weakness and softness extent of the rock masses in the floor determine the magnitude of the floor heave. For example, the surrounding rock masses of the horizontal transporting principal roadway which is located in the Level-780 of the Huainan Xieyi Coal Mine are mainly composed of mudstones and sandy mudstones. The bedding fractures are developed. Compared with the magnitude of the floor heave in the section of the bedded grey fine sandstones, the magnitude of the floor heave in this section is higher for 3 to 4 times;

(3) The thickness of the soft and weak rock strata in the floor also has a significant effect on the floor heave. The thicker of the weak and soft rock strata, the easier for the floor to show the floor heave. The tectonic stress in the stress of the rock masses and the additional stress that is resulted by the mining and tunnelling activities are the main factors in influencing the floor heave of the roadway.

B The stress in the rock masses

(1) For the tectonic stress, the horizontal stress is the main part. On the condition that the horizontal stress is larger, it is adverse for the two sides of the roadway and the floor.

(2) When the roadway is located in the area of the additional stress that is resulted by the mining and tunnelling activities, the roof weighting shows around the surrounding rock masses of the roadway. Displacement occurs towards the direction that is adjacent to the gob area. Furthermore, the rock properties of the floor are bad. Additionally, it is lack of the effective reinforcement and support. The floor heave is severe.

As shown in Figure 5-15(a), the spacing between two layers of the strong rock strata in the top and in the bottom is small. The decreasing velocity of the creep is quick for the thin layer weak

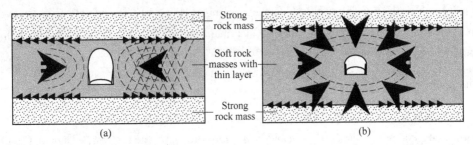

Figure 5-15 The creep evolution process of the surrounding rock masses
when the roadway floor is in the unreinforced state

(a) When the spacing of the strong rock strata is small; (b) When the spacing of the strong rock strata is large

rock. The roadway is easy to stay in the stable state. Therefore, in many coal mines, implementing the "pressure yielding" and the "twice reinforcement" in the weak roadways is significantly effective.

As shown in Figure 5-15(b), the roadway lies in the weak rock with thick layers. The creep displacement process of the surrounding rock masses is not affected by the related restraining force. The long-term creep velocity of the surrounding rock masses of the roadway will not decreased. As a matter of fact, it will accelerate with the loose circle of the surrounding rock masses increasing. The pressure that the loose circle applies on the reinforcement materials will also increase with the loose circle increasing. The reinforcement is difficult and the cross-section of the roadway decreases. Additionally, the floor heave is severe.

5.2.1.3 The function of the water-physical

The water is accumulated in the floor of the roadway. The existing of the water make the floor heave become severer. This is mainly reflected in the following three aspects:

(1) After the rock is immerse in the water, the strength of the rock decreases. This makes the floor become easier to be damaged;

(2) On the condition that the floor is clay stones which are mainly illites and kaolinites, degradation occurs in it after it is immersed in water. Furthermore, it disintegrates and cracks until the strength is completely lost. As a consequence, the floor heave with squeezing and flowing is formed;

(3) When the floor is the expansion-type rock masses such as the montmorillonites, the floor heave with expansion will occur.

5.2.1.4 The supporting and reinforcement parameters

Strengthening the support and reinforcement on the floor has a significant effect in controlling the floor heave of the roadways under the high stress and weak rock mass condition. However, the floor of the roadway is usually in the unsupported or unreinforced state. The main reasons are as following:

(1) In the production process, for the safety concerning, support or the reinforcement is always conducted on the roof of the roadways and two sides of the roadways. This is to prevent the roof collapsing and the rib spalling. It is believed that even failure occurs in the floor, the influence is not significant;

(2) For the output of the gangues when digging the floor, the amount of work is huge. The procedures in casting the bottom arch is complicated;

(3) Implementing the anchorage of the floor is relative more difficult;

(4) Once the support and the reinforcement cannot control the floor heave, the support and the reinforcement that are damaged still need to be cleared when digging the floor. Therefore, the amount of work is larger.

5.2.1.5 The cross-section shape of the roadway

In the mine sites in China, the cross-section shape of the roadway mainly has two types, namely the fold line shape and the curve line shape. The cross-section of the roadway with the curve line shape can have the optimisation effect on the tensile stress of the surrounding area. This makes the tensile stress in the surrounding area of the roadway cross-section decrease. Even the compressive stresses occur, which is beneficial for the stability of the roadway surrounding rock masses.

Excavating the roadway with different cross-section is basically equivalent to excavating the radial roadways with the same diameter. Or it means that there exists a concept of the "Equivalent excavating". The low effective reinforcement zone is the "difference set" between the cross-section of the roadway that is equivalently excavated and the actual cross-section of the roadway. When the shape of the cross-section is different, the low effective reinforcement zone is also different. However, the distribution of the plastic zone is basically consistent, as shown in Table 5-1.

Table 5-1 The appropriate condition of different cross-section shapes

Purpose of the roadway	Name of the cross-section	Application condition
The mining roadway	Rectangle (trapezoid)	The surrounding rock masses are relatively intact and it has the rock bolt implementation condition
	Semi-arch	The surrounding rock masses are fractured. The cracks are extremely developed. The anchorage ability of the rock bolts is poor. There is no rock bolt implementation condition
The development roadway	Semi-circular arch	The stress in the surrounding rock masses is high. The relative difference between the horizontal stress and the vertical stress is small. The deformation of the surrounding rock masses is small
	Ellipse shape (The long axis is along the direction of Y)	The stress in the surrounding rock masses is high. Compared with the vertical stress, the horizontal stress is apparently smaller. The subsidence of the roof is extremely severe
	Three-centred arch	The stress in the surrounding rock masses is high. The horizontal stress is relatively higher than the vertical stress. The deformation of two sides of the roadway is relatively severe. Furthermore, the floor heave is not obvious

Continued Table 5.1

Purpose of the roadway	Name of the cross-section	Application condition
The development roadway	Horseshoe shape	The surrounding rock masses are loose and fractured. The stress in the surrounding rock masses is high. The shrinkage of the cross-section of the roadway is relatively severe, especially the floor heave and he convergence of the two sides of the roadway
	Circular shape	The surrounding rock masses are loose and fractured. The stress in the surrounding rock masses is high. The magnitude of the horizontal stress is equivalent to that of the vertical stress. The convergence of two sides of the roadway and the convergence of the floor are extremely severe
	Ellipse (The long axis is along the direction of X)	The surrounding rock masses are loose and fractured. The stress in the surrounding rock masses is high. The horizontal stress is higher than the vertical stress. The convergence of two sides of the roadway and the convergence of the floor are extremely severe

Then, the influence of the coefficient of the horizontal pressure on the plastic zone of the surrounding rock masses is discussed. On the condition that the vertical stress is equal to the horizontal stress, the cross-section of the equivalent excavating is circle. However, on the condition that the vertical stress is higher than the horizontal stress, the cross-section of the equivalent excavating is ellipse. To be more specific, the long axis of the ellipse is along the vertical direction. On the condition that the vertical stress is smaller than horizontal stress, the cross-section of the equivalent excavating is still the ellipse. However, in this case, the long axis of the ellipse is along the horizontal direction.

The reasonable cross-section profile should consider the stress distribution character, the roadway purpose, the service period and the self-stability of the surrounding rock masses.

5.2.1.6 Classification of the influence extent of the floor heave

According to the different extent of the floor heave, the floor heave can be classified into the following four types, namely the slight floor heave, the apparent floor heave, the severe floor heave and the destructive floor heave.

The slight floor heave indicates that the magnitude of the floor heave ranges from 100mm to 200mm. The heaving of the floor is slow. Therefore, only slight maintaining is needed.

The apparent floor heave indicates that the magnitude of the floor heave ranges from 200mm to 300mm. The variation of the floor is apparent. Therefore, it needs timely maintaining and processing.

The severe floor heave indicates that the magnitude of the floor heave ranges from 300mm to 500mm. The development of the floor heave is rapid. Furthermore, it lasts for a long time, which severely threatens the safety in transporting. As a consequence, all production should be stopped and the floor needs to be dug. The expanding, maintaining and processing should be conducted.

The destructive floor dump indicates that the magnitude of the floor dump ranges from 500mm to

800mm or even higher. The magnitude of the floor dump is large. The destructiveness of the floor, two sides of the roadway and the surrounding rock masses of the floor is large. The loss of the roadway cross-section is more than 1/5. The roadway must be repaired completely and then it can be used.

5.2.2 The controlling theory and technology of the floor heave of the roadways

5.2.2.1 The method of floor brushing

For the roadways which have severe floor heave, the floor brushing can restore the roadway to the original cross-section within the minimum duration. As a matter of fact, it is the simplest and fastest method to solve the problem of the floor heave. However, it is also the method that only cures the symptoms rather than curing the essence. According to the statistics, the floor brushing techniques that are commonly used can be classified to the following three types, based on the instrument that is used to destruct the rocks, namely the method of compressed air pick with manual operation, the drilling and blasting method and the mechanised floor brushing method.

The mechanical dinting method: a machine with multiple functions which can accomplish the dinting procedures individually. It can realise the operations such as crushing, the excavating, the cleaning and smoothing, as shown in Figure 5-16 and Figure 5-17.

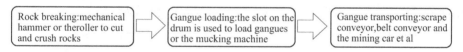

Figure 5-16 The technical flowchart of the mechanical dinting

Figure 5-17 The crushing and loading dinting machine (WPZ-200/60) that is used in coal mines

5.2.2.2 Support and reinforcement strengthening method

The support and reinforcement strengthening method indicates that the support and reinforcement should be strengthen for the rock masses in the floor of the roadways that have the tendency to show the floor heave. The purpose of it is to increase its strength. Therefore, the stability of the rock mass in the floor can be improved. Additionally, the magnitude of the floor heave can be decreased. The principal support and reinforcement strengthening method including the following forms, namely the rock bolts and cable bolts for the bottom floor, grouting in the floor, the closed support and the concrete for preventing the floor heave.

The concrete for preventing the floor heave is a strengthening measure that is applicable for roadways for the permanent use. The hollow slot is excavated in the floor of the roadway, according to the pre-defined depth and shape. The concrete is poured to become the reverse arch for the floor, as shown in Figure 5-18. The advantages of it is that the supporting resistance that is applied on the floor is relatively higher and it is relatively uniform. To enhance the reverse arch of the concrete in the floor, it can also be used together with the metal retractable bottom beam. Therefore, the relatively larger residual deformation resistance for resisting the floor heave can be acquired. The reinforcement effect is shown in Figure 5-19.

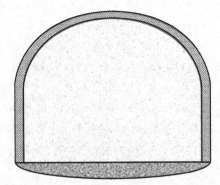

Figure 5-18　The schematic diagram showing the concrete reverse arch

Figure 5-19　The reinforcement effect showing the concrete reverse arch

5.2.2.3　Pressure relieving method

The purpose of the pressure relieving method is to make the rock masses in the floor be located in the pressure relieving zone, through changing the stress state of the surrounding rock masses in the roadway. Therefore, the stable state of the rock masses in the floor can be guaranteed. It is applicable for controlling the floor heave of the roadways under the high in-situ stress and weak rock condition.

In the current stage, the commonly used pressure relieving methods include the following forms, namely the pressure relieving with cutting the crack, pressure relieving with the borehole drilling, pressure relieving with the loosening and blasting, and the pressure relieving with the roadway tunnelling. For the pressure relieving methods to prevent the floor heave of the roadways, they all have their own characters and application conditions. These methods should be selected according to the engineering and geological conditions of the roadways, the type of the floor heave and the related parameters.

5.2.2.4　The joint controlling method

The joint controlling method indicates that those several measures in the reinforcement strengthening methods are combined together. Or the reinforcement strengthening method and the pressure relieving method are fully used. The pressure relieving is conducted firstly. After that, the reinforcement is conducted. Therefore, one is the pressure reliving and the other is the supporting or reinforcing. Double methods are used, which will apparently improve the controlling effect of the floor heave, as shown in Figure 5-20.

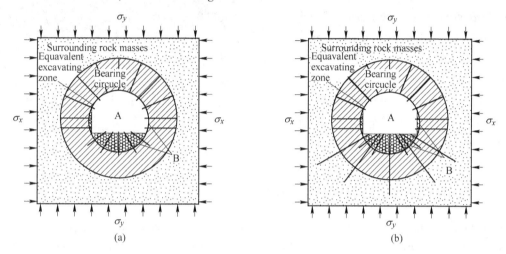

Figure 5-20 The schematic diagram showing the controlling technology for the roadway floor
(a) The floor strengthening technology of normal rock bolts; (b) The integrated technology of locking and grouting for the floor

5.2.3 Materials and equipment used in controlling the roadway floor heave

5.2.3.1 The hollow grouted rock bolts (cable bolts)

The hollow grouted rock bolts are applicable for the complicated engineering and geological conditions, such as the high in-situ stress, the fractured weak and soft rock masses and large deformation. The self-drilling hollow grouted rock boltscan combine the drilling of the rock bolt, the installation, the grouting and the anchoring together, becoming an integral. This prevents the influence of the borehole collapsing on the construction. In particular, it is applicable for solving the problem of the floor heave, as shown in Figure 5-21.

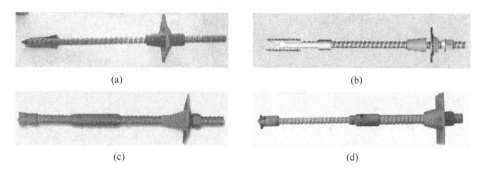

Figure 5-21 Photos to show the hollow grouted rock bolts that are commonly encountered
(a) Normal hollow grouted rock bolts; (b) Expansion-type hollow grouted rock bolts;
(c) Self-drilling hollow grouted rock bolts; (d) Combined hollow grouted rock bolts

For the hollow grouted cable bolts, the grouting with high pressure is used to fully fill the drilled borehole with grouting liquid. It can realise the fully grouted anchorage. When grouting, the grouting liquid develops towards the fractures in the deep area of the borehole. Therefore, the fractured surrounding rock masses can be bonded and the grouting in the deep borehole can be realised. The hollow grouted cable bolt combines the function of normal cable bolts and grouting in the drilled boreholes together, realising the integration of anchoring and grouting, as shown in Figure 5-22.

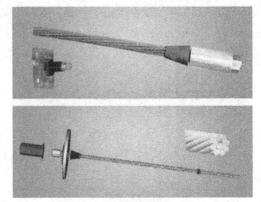

Figure 5-22　Photos to show the hollow grouted cable bolts

5.2.3.2　The pretension cable bolt strand

The pretension cable bolt strand is mainly composed of the metal grouting pipe, anchorage instrument with multiple holes, the plug that is used for stopping the grout, the isolation frame, the guidance cap, several steel strands and the plastic pipe that is used to cover outside. It has the advantages of high elongation rate, high pretension and the integration of anchoring and grouting, as shown in Figure 5-23.

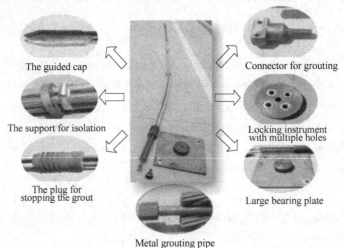

Figure 5-23　Photo showing the pretension cable bolt strand

5.2.3.3　The grouting materials

When selecting the grouting materials, the violence extent of the floor heave in the fractured and weak rock masses, the service life and the strengthening cost should be comprehensively considered. In the roadways that have a long service life and the severe floor heave, the grouting materials that have a high cost can be used, such as the Marithan. On the other hand, in the roadways that have a short service life and slight floor heave, the grouting materials with low cost can be used, such as the micro-fine cement and the soluble glass.

5.2.3.4 The floor borehole drilling equipment

The borehole drilling equipment for the floor is generally divided into the normal drilling machine and the special-purpose drilling machine for the floor. The normal drilling machine has the advantages of universe adapting and low cost. On the other hand, it also has apparent limitation. For example, it is not convenient for implementing and constructing towards the downward direction. Additionally, disposing the gangues after the borehole drilling is difficult. Furthermore, the drilling depth is limited. Last but not least, the time that is consumed is relatively long. Therefore, the special-purpose drilling machine that is used for the floor is especially designed. It has the following advantages.

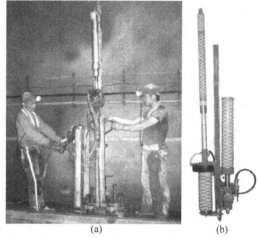

Figure 5-24 The floor borehole drilling equipment
(a) The pneumatic rock bolt drilling machine with trestles (MQJ-120) developed by the Shijiazhuang Aoxun Company;
(b) The supporting drilling machine (ZQJJ120/2.3) developed by the Jiangsu Zhongmei Company

First, the structure supports the roof and the floor, which is convenient for constructing. Secondly, the newly developed dedicated drilling rod is convenient for discharging the gangues. Thirdly, the equipment combines with the drilling rod, drilling the medium deep borehole. Lastly, the drilling machine is efficient and reliable, saving a quantity of time, as shown in Figure 5-24.

5.2.4 Engineering cases of processing the deep mine floor heave

For the roadway floor heave processing, it is a relatively complicated technology. Based on the in-situ roadway stress environment and the surrounding rock mass condition, the specific processing scheme can be determined.

5.2.4.1 The south wing railway roadway floor heave processing case in the Suntong Mine in Huaibei

(1) The maintaining state of the South wing railway roadway.

For the Suntuan South wing railway roadway, the buried depth is 600m and it is influenced by the structures. The effect of the structural stress is apparent. Furthermore, it frequently crosses the coal seams 7# and 8#. For the roadway surrounding rock masses, they are mainly composed of weak and soft rock strata. The stability of the surrounding rock masses is poor. The floor is extremely easy to generate the floor heave. For the roadway cross-section, the width and the height are 5400mm and 4000mm. For the roof, the original support is using rock bolts, support and spraying. Furthermore, the grouting is used to strengthen. The supporting scheme is shown in Figure 5-25. The ground pressure is mainly reflected as that the two sides are squeezed inwards.

The legs of the shed plunges inwards. The top of the shed is cuspidal. The floor heave occurs. The ditch is broken and the clamps are broken. A number of dinting is conducted on the roadway. The accumulated floor heave is up to 1200mm.

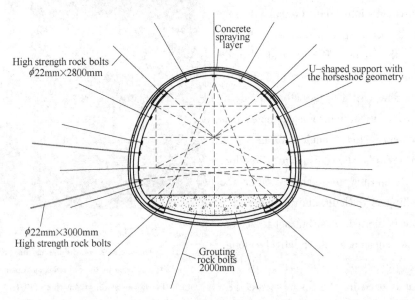

Figure 5-25　The schematic diagram showing the reinforcement scheme

(2) The steel mesh and spraying, the fully closed horseshoe shape support and the grouting controlling technology are used. The technical flowchart of the repairing is shown in Figure 5-26.

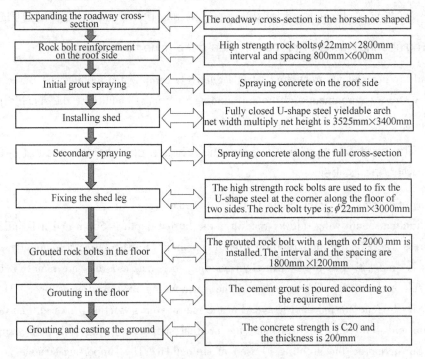

Figure 5-26　The construction technical flowchart of the repairing scheme

5.2.4.2 The other deep mine floor heave controlling engineering case

They are shown in Table 5-2.

Table 5-2 The floor heave controlling technology in the deep mine

Application mine	Controlling technology	Schematic diagram
Taoyuan Mine	Steel mesh and grouting + fully closed horseshoe shaped support + controlling technology with grouting	High strength rock bolts $\phi 22mm \times 2800mm$; Concrete spraying layer; U-shaped support with the horseshoe geometry; $\phi 22mm \times 3000mm$ High strength rock bolts; Grouting rock bolts 2000mm
Xin'an Mine	The controlling technology with fully closed high strength geogrid, arch frame and concrete lining	
Kouzidong Mine	The controlling technology with steel pipes and concrete support	Hollow grouting cable bolts; Cable bolts; High compressive concrete spacer; Compressible steel pipe; Concrete arch support; Layer with spraying concrete; Inverted arch

5.3 The bump and outburst hazard coupling and its prevention technology of the island fully mechanised working face with large mining height and high risks

5.3.1 The island working face bump and outburst coupled parameter studying and analysing

5.3.1.1 The in-situ mining condition of the typical island high hazard fully mechanised working face with large mining height

The working face Wu 9.10-12160 belongs to the exploiting of the island coal pillar. The buried depth is 620.5~736.2m. The mean thickness of the coal seam is 4.3m and the mean dip angle is 10 degrees. The immediate roof is the sandy mudstone, with the mean thickness of 6.5m. The main roof is the fine sandstones, with the mean thickness of 8.2m. When the distance to the coal seam roof is 3.4~8.8m, it is the coal seam Wu 8. The immediate floor is the mudstone and the sandy mudstone. In this area, there are totally 13 faults disclosed. The fall of the faults is 0.4~3.8m. The gas pressure in the coal seam is 1.6MPa. And the gas content is $18m^3/t$. It is designed that the internal section of the working face has a length of 191m and the external section of the working face has a length of 212m. The mining height is 4.3m. There are four roadways that are newly arranged. According to the elevation, from the top to the bottom, the ventilation roadway, the drainage roadway at the top position, the middle roadway at the top position, and the railway roadway are arranged. Additionally, the tail roadway of Wu 9.10-12180 is located above the railway. In the original design of the ventilation roadway, the width of the net coal pillar is 5.85m. In the railway section, the width of the net coal pillar is 10.85m. The arrangement of the working face is shown in Figure 5-27.

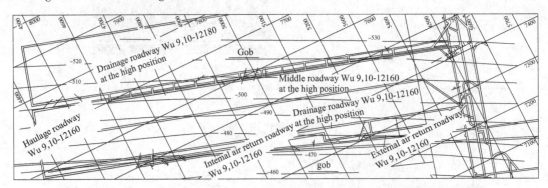

Figure 5-27 The arrangement diagram of the working face Wu 9.10-12160

5.3.1.2 Analysis on the coupled parameters that influence the island working face bump and outburst

(1) The form of the coal pillar for protecting the roadway. For the roadway protecting coal pillar,

the stress concentration is easy to form. Furthermore, it has adverse influence on the overall subsidence of the floor. Therefore, using the tunnelling method that the small coal pillar is used, the abutment pressure shifts towards the deep area of the coal seam. After the roadway is influenced by the retreat mining of the working face in this zone, the deformation of the surrounding rock masses that are located in the overlapped zone of the abutment pressure, will increase prominently. Normally, the maintaining of the roadway is not too difficult.

(2) Mining depth. With the increasing of the mining depth, the probability of the rock burst also increases.

(3) Geological structure. In the middle section and top section of the coal seam Wu 9.10, there are weak coal seam layers. They belong to the non-uniform coal seam which has the coal and gas outburst, and the rock burst. Additionally, during the tunnelling period of the island working face, there are totally 13 faults revealed. The area that is around the faults, is easy to generate rock bursts and outburst.

(4) The coal seam thickness. According to the statistics of the documents, the quantity of the rock bursts in the coal seams with the thickness ranging from 4m to 8m is 6 times larger than the quantity of rock bursts in the coal seams with the thickness ranging from 1m to 2m.

(5) The stress concentration extent. Under the influence of the mining activities, the in-situ stress leads to the micro-fracturing of the coal and the symbiotic rock masses. This creates condition for the gas desorbing and expanding. The high-pressure gases that are desorbed and expanded, become the driving force to form the rock burst. Those two aspects act mutually.

5.3.2 The coupled mechanical model and occurring mechanism of the rock burst and outburst

In the island working face Wu 9.10-12160, there are relatively high in-situ stress and the gas expansion pressure. Under the disturbing of the coal mining and the roadway tunnelling, the induced form of the disaster may be the composite of the solid and the gas energy. However, the form of two kinds of energy cannot be the simple superposition. Therefore, it is essential to establish a comprehensive model that includes two kinds of energy factors. This is to reveal the mechanism of the coupling between the rock burst and the outburst. Its mechanical model is shown in Figure 5-28. The disaster classification and its strategies are shown in Table 5-3.

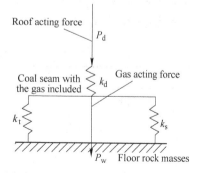

Figure 5-28 The mechanical model that considers the in-situ stress, the mining induced stress and the gas acting force

Table 5-3 The disaster classification and the strategies of the island working face Wu 9.10-12160

No.	Energy condition	Disaster type	Forecasting and predicting methods	Principal solving measures
I	$E_w \propto E_m$ and $E \geqslant U_{min} + W_{min}$	Rock burst	1. Comprehensive index method 2. Drilling method 3. Eletro-magnetic radiation method	1. Water injecting in the coal seam 2. Pressure relieving with drilling holes 3. Pressure relieving with blasting
II	$E_w \propto E_m$ and $E_w \geqslant U_{min} + W_{min}$	Coal and gas outburst	1. Comprehensive index method 2. Drilling method 3. Gas emission initial velocity method with drilling hole 4. Eletro-magnetic radiation method	1. Coal seam gas drainage 2. Water injecting in the coal seam 3. Advancing hole drilling 4. Loosing and blasting in the shallow borehole
III	$E_w \cong E_m$ and $E_f \geqslant U_{min} + W_{min}$	Composite disaster of coupling between bump and outburst	I and II	I and II

5.3.3 Numerical simulation of the stress field in the island section and the proper position of the horizontal roadway

5.3.3.1 The numerical calculation model and simulation scheme of FLAC

The working face Wu 9.10-12160 belongs to the exploiting of the island coal seam. The elevation is ranged from −470.5m to −526.2m. The buried depth is 620.5 ~ 736.2m. For the island working face Wu 9.10-12160, the original design requires that the width of the coal pillar is 5.85m. For the railway roadway, the width of the coal pillar is 10.85m. The horizontal distance between the middle of the internal section of the ventilation roadway and the middle of the drainage roadway at the top position is 15m. As for the horizontal distance between the middle of the external section and the middle of the drainage roadway at the top position is 35m. The external section of the conveyor roadway is the influencing section of the externally-misaligned gas drainage roadway of the working face Wu 9.10-12180. The internal section is the influencing section of the gas drainage roadway that is not externally-misaligned. Its simulation scheme is shown in Table 5-4.

Table 5-4 Simulation scheme

	Type		Width of the coal pillar		Type		Width of the coal pillar
Ventilation roadway pillar	Internal section		3m	Haulage roadway pillar	Internal section		3m
			4m				8m
			5m				11m
			6m				14m
	External section		3m		External section		8m
			4m				10m
			6m				11m
			8m				12m

5.3.3.2 The value calculation and the optimisation determination of the remaining of the ventilation roadway coal pillar

The simulation results are shown in Figure 5-29 and Figure 5-30. When the coal pillar with the width of 4m is reserved between the ventilation roadway and the gob area, the stress that is suffered by the coal pillar is relatively large at the beginning stage. However, after the deformation and failure occur in the coal pillar and the coal pillar is compressed to failure, the suffered stress decreases rapidly to be lower than the in-situ stress. When the coal pillar width is ranged from 5m to 10m or even larger, the elastic core occurs in the coal pillar. Furthermore, the relatively large elastic energy is accumulated because of the relatively large in-situ stress. Therefore, from the perspective of prevention of the rock burst and outburst hazard, decreasing the self-igniting in the gob area and reducing the waste of the coal resources, the ventilation roadway coal pillar width should be 4m.

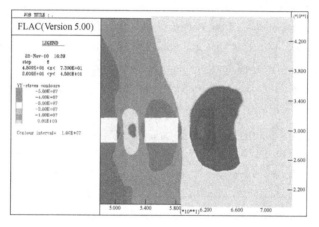

Figure 5-29 The vertical stress distribution diagram of the roadway surrounding rock masses when the coal pillar with a width of 4m is set for the internal area of the ventilation roadway

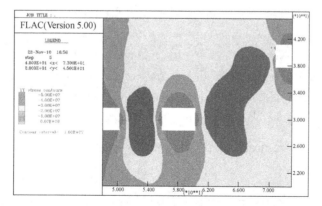

Figure 5-30 The vertical stress distribution diagram of the roadway surrounding rock masses when the coal pillar with a width of 6m is set for the internal area of the ventilation roadway

Through conducting the simulation and the in-situ practical construction, when the adjacent externally-misaligned gas drainage roadway is located above the haulage roadway of this zone, the width of the remained coal pillar between the haulage roadway and the gob area should be 10.85m. Under this condition, the outburst and the rock burst controlling effect is relatively better.

When there is no adjacent working face, for the haulage roadway section of the externally-misaligned gas drainage roadway (around 210m), due to the fact that there is no pressure relieving effect of the drainage roadway, in the exploiting design, the coal pillar width is remained unchangeable. Therefore, the in-situ stress is relatively higher. Also, the risk of the rock burst and the outburst is relatively larger. The roadway tunnelling velocity decreases. In the in-situ construction process, the measures to prevent the outburst and rock burst are strengthened.

5.3.4 Experimental study on the bump tendency of the coal and rock masses in the is land working face

The measured results of the coal samples, and the composite coal and rock samples are shown in Table 5-5. From the experiment and the calculated results, it can be acquired that the coal sample of the island working face Wu 9.10 has the weak burst tendency. The composite coal and rock samples of the coal seam Wu 9.10 and the mudstones in the roof and floor, have the weak burst tendency. It can be known that the coal and rock samples that are from the island working face Wu 9.10-12160 in the Pingba Mine, have the weak burst tendency.

Table 5-5 The statistics showing the average value of the measured results for the burst tendency parameters of the coal samples, and the composite coal and rock samples

Sample type	Dynamic failure time/ms	Burst index (KE)	Elastic energy index (WET)	Burst tendency results
Pure coal sample	466	2.14	9.13	Weak burst
Composite coal and rock sample	314	1.54	4.25	Weak burst

Although the coal and rock samples from the island working face Wu 9.10-12160 have the weak burst tendency, due to the fact that the in-situ stress and the gas pressure of this island working face are both high, under the mining influence, the risk that the rock burst and the outburst, and the coupled composite mine disasters will occur, is still pretty high. Therefore, during the working face exploiting process, the drilling method, the electro-magnetic radiation method, and the composite index method will still be needed to predict. The disaster prevention measures such as the water injection in the coal seam, the blasting with pressure relieving and the gas drainage, will be adopted.

5.3.5 The bump hazard grading and the partition predicting of the mining zone with the large mining height

5.3.5.1 The partition prediction of the bump hazard for the island working face Wu 9.10-12160

For the island working face Wu 9.10-12160, in the whole area, the structure of the roof and the floor, and the basic physical properties of the coal seam will not change significantly. However, the distribution of the faults in the working face and the dimension of the coal pillar in the different areas of the working face are different.

According to the distribution of the faults and the dimension of the coal pillars, the working face are divided into three sections, namely A, B and C.

The section A indicates the area from the open-off cut to the place that is 130m far away from the open-off cut.

The section B indicates the area from the place that is 130m far away from the open-off cut to the place that is 330m far away from the open-off cut. This section has a length of 200m.

The section C indicates the area from the place that is 330m far away from the open-off cut to the designed stopping line.

According to the geological conditions and the mining conditions of the working face Wu 9.10-12160, the rock burst hazard grade is predicted. And the results are that the sections A and C belong to the medium rock burst hazard zone. The section B belongs to the medium or even higher hazard zone, as shown in Figure 5-31.

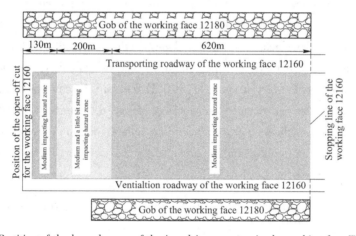

Figure 5-31 Partition of the hazard zones of the impulsion pressure in the working face Wu 9.10-12160

The core zone of the rock burst prevention is the zone of 200m from the position that is 130m far away from the open-off cut to the position that is 330m far away from the open-off cut.

In this section, the gob with two sides is transited to the gob with three sides. In the transition section of the ventilation roadway, the stress concentration will occur and the pressure will be relatively larger. Meanwhile, in this area, there are faults F12 and F13. Among them, the fall of the fault F12 is 3.8m, which will have a relatively big influence on the retreat mining of the working face.

5.3.5.2 The roadway deformation observation of the island working face Wu 9.10-12160

To effectively predict the rock burst, the roadway deformation observation is conducted on the rock burst hazard area. In the entity coal section along two sides of the ventilation roadway in the island working face Wu 9.10-12160, two observation stations are arranged. Also, two observation stations are set along the gob section side of the ventilation roadway, as shown in Figure 5-32.

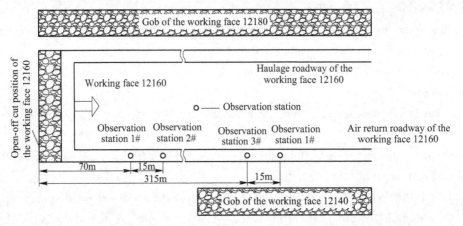

Figure 5-32 The roadway deformation observation station layout diagram

After calculation, it is found that in the observation period, for the entity coal section of the ventilation roadway, the mean convergence of the roadway two sides is 6.0cm. The convergence velocity is 0.222cm/d. The mean convergence between the roof and the floor is 7.5cm. The convergence velocity is 0.278cm/d. For the ventilation roadway gob area, the mean convergence of two sides is 5.0cm. The convergence velocity is 0.185cm/d. The convergence between the roof and the floor is 7.0cm. The convergence velocity is 0.259cm/d. The roof and the floor deforming velocity of the ventilation roadway, and the deforming velocity of two sides are in the reasonable range. This indicates that in the observation period, the stress variation is gentle. And, there is no sudden appearance character of the rock burst.

5.3.5.3 The electromagnetic radiation and the drilling monitoring and the evaluation index

The dynamic hazards of coal masses and rock masses, such as the outburst of coal and gases, and the rock burst, are the mutual acting results of the mutation of the in-situ stress and other factors. Before it occurs, there are apparent character laws. Therefore, the electro-magnetic radiation method can be used to conduct the prediction on the dynamic hazard of the coal masses and rock masses, such as the outburst and the rock burst. Also, the drilling method can be used. According to the variation law of the discharged coal ash and the corresponding dynamic effect, the prediction can be conducted based on the distinguishing of the burst hazard.

The observation principle of the drilling method is that the quantity of the drilling that is discharged in the borehole drilling construction comprehensively reflects the factors in terms of three aspects, namely the stress state of the coal seam, the mechanical character of the coal and the gas, to a certain degree. During the borehole drilling construction, the released potential energy, by the methods of the coal mass displacement, squeezing, friction and fracturing, is reflected with the form of the increment of the drilling. Under the same borehole drilling construction technique, the larger the coal seam stress, the higher the gas pressure, the smaller the coal strength. Also, the generated quantity of the drilling is larger. This indirectly reflects that the risk of the rock burst and the outburst is larger.

Specifically, when the power discharging rate with a unit length increases or even beyond the designated value, it indicates that the stress concentration extent increases and the burst hazard increases. The corresponding theory indicates that for the high-gas coal seam, the drilling index for the prediction of the rock burst and outburst hazard, it can be expressed with the drilling quantity with the unit length:

$$S = \rho_c \pi (\eta a)^2 \left\{ 1 + \frac{\sqrt{3}\sigma_c}{E} \left[1 + \frac{E}{\lambda} - (q-1) \frac{E}{\lambda} \frac{aP_i}{\sigma_c} \right] \right\} \quad (5-2)$$

$$q = \frac{1 + \sin\phi}{1 - \sin\phi} \quad (5-3)$$

5.3.6 The prevention scheme and parameters of the bump and outburst

Through conducting the theoretical analysis, laboratory research, mechanical modelling and in-situ observation on the burst and outburst coupling law of the working face Wu 9.10-12160, the basic parameters such as the gas drainage (the spacing between the drilled boreholes is 2.01m and the drainage period is more than 140 days), and the waster injecting in the coal seam (the length of the drilled borehole is more than 8.5m) are calculated. Furthermore, the comprehensive processing scheme in which the gas drainage, the water injection in the coal seam and the blasting with loosening are integrated, is formed and optimised.

5.3.6.1 The scheme design of the gas drainage

For the fully mechanised working face with large mining height (Wu 9.10-12160), the comprehensive drainage method in which the drainage with the drilled boreholes that cross the layers, the drainage with the drilling boreholes in this coal seam, and the drainage with the shallow boreholes in the working face are integrated. The drilled borehole layout is shown in Figure 5-33.

(1) The layout form of the drilled boreholes in this roadway of this coal seam.

In the haulage roadway, the original construction for the drilled boreholes requires that the mean depth is 52m and the borehole diameter is 113mm. The borehole spacing is 4m. Then, the borehole spacing is decreased to 2m and the mean borehole depth is 105m. In the haulage roadway construction, there are 787 drainage boreholes. The total depth of the borehole is 68590m. In the ventilation roadway, there are totally 380 gas drainage boreholes. The borehole diameter is 113mm and the mean borehole depth is 94mm. The mean borehole spacing is 2.7m. The total borehole depth is 35720m. The drainage time is more than 180 days. In the working face, the coverage rate of the drilled boreholes reaches 100%.

(2) The layout of the shallow drainage boreholes in the working face in this coal seam.

In the range of 145m that is below the ventilation roadway for 20m and above the haulage roadway for 20m, there are two rows of drilled boreholes whose diameter is 89mm. The borehole spacing is 1.5m. For the first row of boreholes, the borehole depth is 10m. For the second row of boreholes, the borehole depth is 16.5m. In each row, there are 98 boreholes. For the last borehole, the networking drainage time is more than 2 hours.

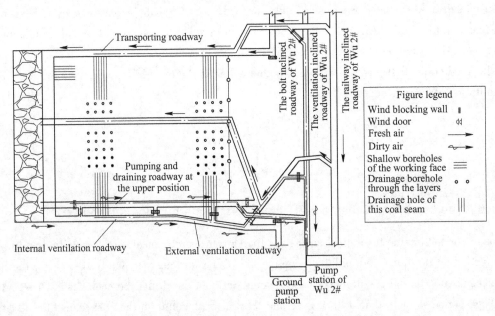

Figure 5-33　The schematic diagram showing the layout of the drilled boreholes to drain the gases in the working face Wu 9.10-12160

5.3.6.2　The water injection scheme in the shallow borehole in the working face

In the mine 8#, the water injecting in the shallow borehole at the working face is used to processing the outburst and rock burst.

(1) The depth of the water injecting hole.

If the depth of the water injecting hole crosses the stress concentration zone, the effect is best. The depth of the drilled borehole should be more than 8.5m.

(2) The water injecting hole spacing.

The spacing of the water injecting hole is determined by the wetting radius of the coal seam. Apparently, the larger the wetting radius, the larger the water injecting hole spacing. To make sure that the coal seam between the two water injecting holes can be fully wet, A1 should be at least less than 2 times of A2. A1 indicates the area of the rectangle ABCD. A2 indicates the diffusing area of the injected water in each water injecting hole in the rectangle ABCD.

The schematic diagram of the water injecting layout is shown in Figure 5-34. After calculation, the spacing between the water injecting hole is: $S = 1.54$m. In the working face, the water injecting hole selects 1.5m.

(3) Water injecting pressure.

Considering the water injecting equipment in the shallow boreholes in the working face and the borehole sealing, 3~5MPa is selected.

(4) The quantity of the moist coal that is born by each borehole.

When the water injecting is conducted in the shallow boreholes in the coal face, the quantity of the moist coal that is born by each hole is 69.7t.

(5) The quantity of the injected water in each hole is 3.8m³.

(6) The anti-bump and anti-burst scheme with water injecting in the mine 8#.

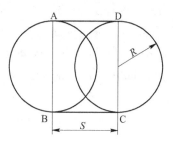

Figure 5-34 The water injecting layout diagram

In the range of 145m in the middle of the working face, two rows of boreholes with the diameter of 89mm and depth of 16.5m are set directly. These are advancing exploring, pressure relieving, drainage and water injecting holes. The hole spacing is 1.5m. After each hole is drilled, the networking drainage is conducted. After the drainage, the water injection is conducted in the coal seam. This is conducted until the coal face leaks water or the adjacent borehole leaks water.

5.3.6.3 The loosening and blasting scheme of the working face

After the working face Wu 9.10-12160 performs the local comprehensive processing measures, in the middle of the working face that is below the ventilation roadway and above the haulage roadway for 20m, the vibration blasting borehole with a diameter of 89mm and depth of 16.5m is set for every 10m.

The loosening and blasting hole:

On the first day, 8 holes are blasted. The loosening and blasting holes are set in the hydraulic supports 14#, 28#, 42#, 56#, 70#, 84#, 98# and 112#;

On the second day, 8 holes are blasted. The loosening and blasting holes are set in the hydraulic supports 21#, 35#, 49#, 63#, 77#, 91#, 105# and 119#;

On the third day, no blasting is conducted. Three days constitute a cycle. For each cycle, the required footage is 6m. The loosening and blasting advancing distance of 10.5m is retained.

5.3.6.4 Engineering effects

According to the underground observed data statistics, the predicted results of the electromagnetic radiation, the amount of the drilling, and the gas emission velocity of the working face Wu 9.10-12160 have the consistence. At the same time when the current amount of the drilling and the gas emission velocity monitoring are conducted, together with the monitoring of the electromagnetic radiation, the prediction and the forecasting of the outburst of the working face Wu 9.10-12160 and the rock burst can be realised. Based on the theoretical calculation and the in-situ measuring, it is determined that in the retreat mining period of the working face, for the critical index of the amount of the drilling, the currently used value is 5.0kg/m. It is determined that the warning value of the electromagnetic radiation amplitude is 58mV. For safety concern, 1029 is used for the pulse value.

Through implementing the research outcome and technical scheme of this project, the cost on the outburst prevention and rock burst reducing is decreased. The resource recovery rate is improved. The safety benefits are obvious. After the working face starts production, the safety production is always maintained. This improves the safety exploiting condition of the working face. The accidents of the personal injury are eliminated. Furthermore, the heavy labour strength which is induced by the operators' processing of the outburst and the rock burst accidents, is eliminated.

5.4 The disaster mechanism and adjusting of the rock burst and underground reservoir[17,18]

The rock burst means that under the condition of the excavation and the disturbing of the other outside environment, the elastic deformation potential energy that is accumulated in the underground engineering rock masses, releases suddenly. This leads to the dynamic phenomenon, such as the splitting and injection of the surrounding rock masses. The rock burst usually occurs in the construction engineering, such as the water conservancy and hydropower, the transportation, and metal mines in the deep area. It is the worldwide difficult issues in the rock mechanics area. It has pretty strong burstiness, randomness and hazard.

5.4.1 The variation law and mechanism of the rock bursts with different types

From the space-time appearance character, the rock burst can be divided into the immediate rock burst and the time-lag rock burst. From the mechanism, it can be divided into the strain rock burst and the strain-structural surface slippage rock burst. The developing mechanism of the rock burst means the corresponding mechanical mechanism variation law during the rock masses fracture in the developing process (such as the tension fracture, the shearing fracture and the composite fracture). The classification of the rock burst is shown in Figure 5-35.

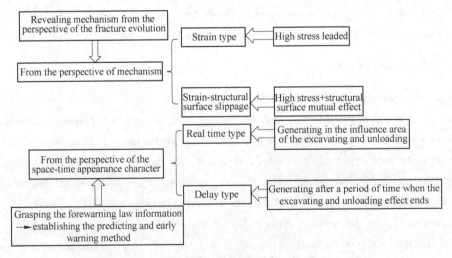

Figure 5-35 The rock burst classification

5.4.1.1 The developing law and mechanism of the immediate rock burst

For the majority of the immediate rock bursts, the microseism incidents increase and the space is accumulated. Before the rock burst occurs, there is no calmness period. The severer, the rock burst, the larger, the number of the micro-fracture incidents. Furthermore, the larger, the energy. The energy releasing shows the high position. Also, it has the decreasing and then increasing trend. The non-elastic volumetric deformation increases continuously. Furthermore, it has the sudden increasing tendency. The immediate rock burst generally includes the immediate strain rock burst and the immediate strain-structural surface slippage rock burst. The statistics of the microquake incidents in the immediate rock burst period, is shown in Figure 5-36.

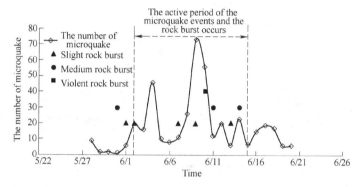

Figure 5-36   Statistics on the microquake incidents in the immediate rock burst period

The developing mechanism of the immediate strain rock burst is mainly composed of tensile failure. As for the developing mechanism of the immediate strain-structural surface slippage rock burst, there are a large number of tensile fracture and occasional shear fracture/composite fracture. For most fracture surfaces, the intersection angle between the strike direction and the axis of the tunnel is less than 40 degrees. The combined cutting forms the blasting crater with the "V" type. In the rock bursting period, the relationship between the energy releasing and the non-elastic volumetric deformation is shown in Figure 5-37.

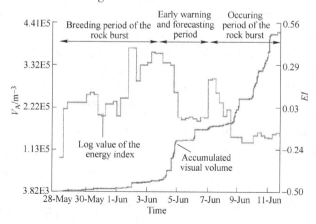

Figure 5-37   The relationship between the energy releasing and the
non-elastic volumetric deformation in the rock burst period

5.4.1.2 The developing law and mechanism of the time-lag rock burst

For the time-lag rock burst, in the excavating process and after the excavation of the rock burst area, the microquake incidents show increasing and gathering. However, before the rock burst occurring, there is an apparent silent period. The non-elastic volume deformation continuously increases. Meanwhile, it has the abrupt increasing tendency. The energy releasing is in the relatively high level and then it has the decreasing tendency. However, before the time-lag rock burst occurs, the variation is relatively small, as shown in Figure 5-38 and Figure 5-39.

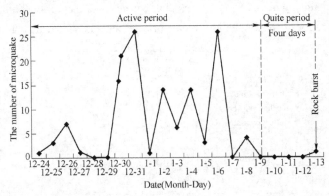

Figure 5-38 Statistics on the microquake incidents before the time-lag rock burst occurs

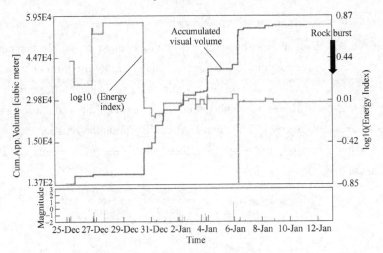

Figure 5-39 The relationship between the energy releasing of the immediate rock burst and the non-elastic volume deformation law

The developing mechanism of the time-lag rock burst: A quantity of tension, shearing and the composite fractures occur commutatively → silent period → shearing fracture → occurring of the rock burst. The in-situ state of the time-lag rock burst is shown in Figure 5-40.

5.4.2 Hazard estimation and predicting of the rock burst

The rock burst hazard estimation and the forecasting include the rock burst hazard estimation before the excavation. In the excavating process of the underground engineering, according to the real-time

5.4 The disaster mechanism and adjusting of the rock burst and underground reservoir

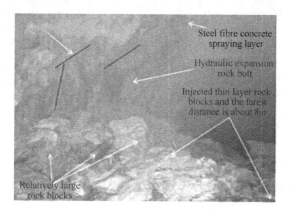

Figure 5-40   The developing mechanism of the time-lag rock burst

monitoring information of the microquake, the warning on the rock burst is conducted. The quantification of the rock burst hazard estimation and the prediction mainly include the following contents:

(1) Classification of the rock burst grade.

Based on the microquake energy threshold and the *SLI* grade section, the rock burst grade can be divided into the slight rock burst, the medium rock burst, the violent rock burst and the extremely strong rock burst. Among them, the *SLI* grade section with multiple indexes, is comprehensively determined by the sound character, the surrounding rock mass fracture character, the failure depth, the reinforcement failure extent and the engineering influencing extent.

(2) The *RVI* index in the rock burst hazard estimation.

The *RVI* index for the rock burst hazard evaluation can be determined by the following equation:

$$RVI = F_s F_r F_m F_g \tag{5-4}$$

Where, F_s is the stress controlling factor; F_r is the rock character factor; F_m is the rock mass system stiffness factor; F_g is the geological structure factor.

Based on the *RVI* index for the rock burst hazard evaluation, the calculating equation for the rock burst blasting crater depth D_f can be acquired:

$$D_f = R_f(C_1 RVI + C_2) \tag{5-5}$$

Where, R_f is the hydraulic radius.

According to the above equation, before the engineering excavation, the estimation is conducted on the rock burst blasting crater depth. Meanwhile, in the engineering excavating process, based on the practical geological condition that is revealed by the excavation, the necessary updated evaluation can be conducted on the rock burst blasting crater depth.

(3) The numerical analysis method on the rock burst hazard estimation.

The numerical analysis method for the rock burst hazard evaluation is based on the mechanical model and criteria. On one aspect, the failure proximity is used to evaluate the failure extent of the rock masses in different positions. On the other aspect, according to the local energy releasing rate, the magnitude of the energy releasing and the total released energy in this area when the rock masses at different position fail can be evaluated.

(4) The rock burst warning probability method of the development law based on the microquake

information.

Nowadays, in this area, the rock burst grade and its probability warning can be acquired by the following equation:

$$P_i = \sum_{j=1}^{6} w_j p_{ji} \tag{5-6}$$

Where, i is the rock burst grade: violent, medium, slight, and none; j is the microquake information: number of incidents, energy and visual volume, incident rate, energy velocity and visual volume rate; p_{ji} is the probability of the microquake information j; w_j is the weight coefficient.

(5) The neural network comparison method of the engineering case for the rock burst hazard estimation.

For the rock burst hazard evaluation engineering case, the neutral network comparison method is shown in Figure 5-41. According to the above contents, the hazard evaluation and the forecasting method on the rock bursts before the excavation and during the excavation process, are conducted. The following comprehensive integrated table, as shown in Table 5-6, is established.

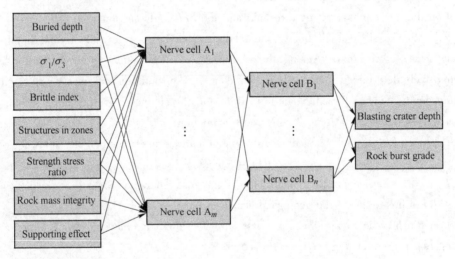

Figure 5-41 The neural network comparison method of the engineering case

Table 5-6 The rock burst hazard evaluation and the comprehensive integrated method of the prediction

The rock burst evaluation method before excavation	The rock burst prediction method during the excavation process	Expression of the results
RVI index method		Depth of the blasting crater, the rock burst grade based on the blasting crater depth
The neutral network predicting method based on the engineering case comparison		Rock burst grade and depth of the blasting crater
The numerical analysis method based on the new indexes such as the local energy releasing rate and the failure proximity		The position and depth of the cross-section where the rock burst hazard is high
	The probability method based on the practically measured microquake information evaluation	The present rock burst grade and probability in this zone

5.4.3 The dynamic adjusting and controlling method of the rock burst developing process

The dynamic adjusting and controlling method of the rock burst developing process mainly includes the "three steps" strategy of the adjusting and controlling of the rock burst developing process, and the hydraulic adjusting and controlling method based on the microquake information evaluation law. The specific hydraulic adjusting and controlling flow chart can be shown in the Figure 5-42.

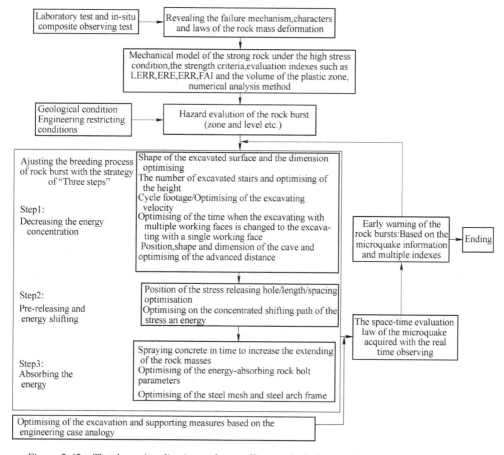

Figure 5-42 The dynamic adjusting and controlling method of the rock burst developing process

According to the above-mentioned dynamic adjusting and controlling method of the rock burst developing process, the excavation and supporting strategy table for preventing the rock bursts with different grades, is established. The specific strategy is shown in Table 5-7.

5.4.4 Practices on the deep buried diversion tunnel and the water discharging tunnel in the Jinping Level two hydropower station

For the Jinping Level two hydropower station, the maximum buried depth of the diversion tunnel and the drainage hole is 2525m. The length of a single hole is 16.7km. 80% of them are the brittle marble. In November 2009, the extremely rock burst occurred in the drainage hole, as shown in Figure 5-43.

Table 5-7 The excavation and support strategy for preventing rock bursts with different grades

Rock burst grade	Geological surveying strategy	Excavation and monitoring warning strategy	Support strategy
Extremely strong rock burst	1. Grasping the rupture, strong structural surface profile and its relationship with engineering 2. Grasping the local geological structure abnormity (axis of the anticline, wing)	1. Dimension optimization. Dime-nsion optimisation of the cross-section profile; Tunnelling advancing optimisation; When tunnelling along the same direction is conducted, before it is cut through, the method is changed to tunnelling with only one working face 2. Special methods. Stress releasing borehole; Stress releasing of the controlling structural surface; Optimisation of the guided hole 3. Monitoring and warning of the dynamic microquake	1. Reinforcement form and parameters (the designed absorbed energy is around $50 kJ/m^2$). Immediate spraying and the required absorbed energy is $60 kJ/m^2$. The energy absorbing rock bolts are installed systematically. The steel bearing plate is added. The required absorbed energy is $39.1 kJ/m^2$. The rock bolt should intersect with the controlling structural surface with large angle. Hanging the mesh and installing the steel arch. Repeated spraying 2. The construction procedures of the drilling and blasting support: Immediate spraying, arrangement of the energy absorbing rock bolts, hanging the mesh, installing the steel arch and repeated spraying 3. Construction procedures of the TMB support: In the zone L1, immediate spraying, mesh hanging, arrangement of the energy absorbing rock bolts, installing the steel arch and the repeated spraying in the zone L2
Violent rock burst			1. Reinforcement form and parameters (the designed absorbed energy is about $22 \sim 50 kJ/m^2$) 2. Immediate spraying and the required absorbed energy is $10.9 kJ/m^2$. It is required that the rock bolt system absorbs energy: $11.1 \sim 39.1 kJ/m^2$. For the others, they are same as the extremely strong rock bursts
Medium rock burst		1. Dimension optimisation. Tunnelling footage optimisation 2. Excavating method (Drilling and blasting method, or the TBM) and cross-section profile, dimension, the construction without rock blasting based on the same condition 3. Real time microquake monitoring and warning	1. The whole designed anti-burst energy: $13 \sim 22 kJ/m^2$ 2. The immediate spraying. It is required that the absorbed energy is $4.7 \sim 10.9 kJ/m^2$. Following the system, the rock bolts are arranged and the steel bearing plate is added. It is required that the absorbed energy is $8.3 \sim 17.3 kJ/m^2$
Slight rock burst		Tunnelling velocity, excavating method (Drilling and blasting method, or the TBM) and cross-section profile, dimension, the construction without rock blasting based on the same condition	1. The whole designed anti-burst energy$< 13 kJ/m^2$ 2. Immediate spraying 3. The random rock bolts and the steel bearing pate are added following the stress concentration zone and the frequent occurring zone of the rock bursts

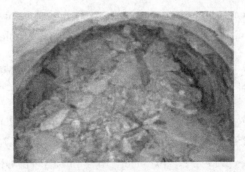

Figure 5-43 Extremely strong rock burst in the drainage hole

Based on the specific geological situation of the diversion tunnel and the drainage hole of the Jinping Level two hydropower station, the following addressing schemes are proposed:

(1) In the violent and the extremely violent section of the hole that is excavated with the TBM, the guided hole processing is conducted. On the aspect of preventing the rock burst with the structural surface, the top guided hole method has more advantages than the middle-guided hole method.

(2) The depth and the position of the stress relieving hole in the extremely violent rock burst hole is optimised. This is beneficial for releasing the energy in advance. The stress concentration section shifts to the deep area of the surrounding rock masses.

(3) The reinforcement parameters in the rock burst hole section with different levels are optimised. The comprehensive absorbed energy of the supporting system is quite difficult to be over than $50kJ/m^2$. When the impact energy of the rock burst is more than $50kJ/m^2$, the excavation strategy should be combined to solve the rock burst hole section.

(4) The rock bolt length should be optimised according to the rock burst with different grades.

(5) The forecasting, early warning and the dynamic adjusting are conducted on the rock burst. In the continuous monitoring period of the micro-quake, 275 rock bursts occur. It is predicted that there are 243 rock bursts, accounting for 88.36% of the number of the practically occurred rock bursts.

The predicting and early warning of rock bursts with different grades are conducted on 241 hole sections (the accumulated length of the hole is 7605m). This avoids the occurring of the rock bursts with different grades of 135 extremely violent rock burst hole section. Also, this decreases the strength of the occurring of the rock bursts with different grades of 13 hole sections (the accumulated length of the hole is 418m). The micro-quake monitoring and the rock burst prediction have a significant effect on the rock burst prevention and the construction guidance. In the kinds of the monitoring of the construction process, there is no serious consequence that is resulted by the rock burst. It ensures the safety of the staff and the equipment in the construction process. The rock burst predicting effect is shown in the above Figure 5-44.

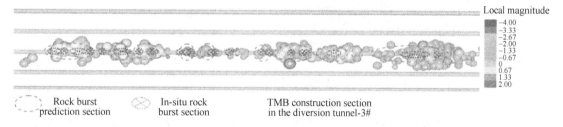

Figure 5-44 The predicting and early warning effect of the rock burst

5.4.5 The disaster mechanism and the prevention theory of the underground reservoir group

5.4.5.1 The research background of the disaster mechanism and the prevention theory of the underground reservoir group

The energy reserve indicates that the petroleum and the natural gases are stored in the large-scale underground chambers and the surface storage tanks. This is to process the influence that is induced by the interruption of the energy importing under the extreme condition (such as the war, the earthquake, the terrorist attack and the extreme environment).

The underground storage of the salt and oil indicates that the large-scale underground space after brining of the salt mine is used to implement the gas storage, as shown in Figure 5-45. For the forming of this storage chamber, the technological process is that the fresh water is injected. And then, the salt mine is corroded. Then, the brine is discharged, the large-scale underground space is formed. Compared with other storage method, the salt rock gas-storage has the prominent advantages such as the high injecting and exploiting efficiency and the small amount of the bottom. It is the internationally recognised suitable place for the storage of petroleum and natural gases. However, in China, the salt rocks mainly belong to the layered structure which is formed by the lacustrine deposit. It has the characters of the thin salt layer, multiple interlayers and shallow buried depth. This character makes that the underground oil gas storage in China is confronted with vital technical problems and challenges. Based on this vital problem, the following four scientific problems need to be solved:

(1) The accurate detecting method and theory of the group of structure;
(2) The constitutive law and failure mechanism of the layered salt rocks;
(3) Mutual acting mechanism and the space-time process of the disaster for the chamber group;
(4) Disaster-causing factors, risk evaluation and the prevention theory.

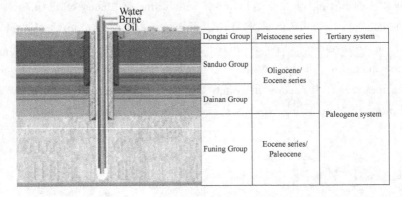

Figure 5-45 The underground storage of the salt and oil

5.4.5.2 The research progress on the disaster theory and prevention theory of the underground reservoir group in China

Under the coupling effect of multiple fields such as the stress field, the seepage field and the oil gas pressure, the deformation law and the strength criteria of the layered salt rocks are the core in the design of the underground chamber group, the analysis of the oil gas leakage and the stability evaluation. Therefore, under the coupling effect of multiple fields, the constitutive law and the deformation failure mechanism of the storage chamber rock masses are the precondition and basement in the research on the disaster prevention.

Among them, the microscomic damaging constitutive law of the layered salt rocks under the triaxial condition, relatively describes the mechanical behaviour of the pure salt rock and the salt rock with the interlayer embedded under different confining pressure. The constitutive law of the Cosserat-like for the bedded salt rocks that concern the interlayer effect, is considered. Then, the new constitutive law of the composite salt rocks that can reflect the interlayer effect is proposed. This provides a new idea and method for the analysis of the deformation and failure of the interlayered salt rocks. The new constitutive law of the creeping of the salt rocks: the derivative model of the grade, is proposed. Through changing the derivative order, the rheology curve can reflect the non-linear variation process of the salt rock strain.

Based on the above-mentioned constitutive law and deformation failure mechanism of the surrounding rock masses in the storeroom, the scholars in China conducted deep research on the surface subsidence of the underground chambers, the penetrating disaster of the layered salt rocks, the stability of the dense storehouse group and the disaster mechanism of the storehouse. A number of outcomes have been acquired.

(1) The surface subsidence mechanism of the storeroom group.

The influence parameters of the surface subsidence of the storeroom area can be listed as the number of cavities, the shape of the cavity, the spacing between the cavity, the operation period and the pressure in the cavity. On the other aspect, the surface subsidence is mainly resulted by the storeroom convergence. Based on this, the functional relationship between the convergence and the subsidence is established. The transfer function method is used to predict the surface subsidence.

(2) The penetrating disaster mechanism of the layered salt rocks.

Based on the migrating law of the gases on the salt rock interlayer interface in the layered salt rocks, the corresponding penetrating model is established. The penetrating migrating character and its range of the gases in the long-term operating process are acquired.

(3) The stability of the dense storeroom group.

To reveal the complicated mechanism of the linked failure of the salt rock storeroom, the systematic research is conducted on the three aspects, including the constitutive simulation of the salt rock material, the stability analysis of the structure and the failure experiments of the chamber group. The concept of the "restrained balance state" in the unbalanced thermal mechanics is regarded as the basement. The stability theory—deformation stability theory of the layered salt rock storeroom that considers the unbalanced state is proposed. Based on the above-mentioned results, the combined analysis process of the integrated stability of the storeroom group, the interlocking failure and the space-time evolution is preliminarily realised.

(4) The disaster mechanism of the storeroom during the cavity construction process.

The similarity theory of the flow field during the cavity construction process is established. The mutual acting mechanism of the interlayer and the flow field during the cavity construction process is revealed. The influencing law of the flow field on the developing of the cavity shape, during the cavity construction period, is preliminarily revealed. For the technical problem of the interlayer collapsing that is encountered during the cavity construction process, the mechanical mechanism of the interlayer collapsing is revealed. This provides theoretical support for the cavity construction controlling technique.

(5) The disaster hazard evaluation and prevention of the storeroom.

Through engineering comparison, the surveying of the specialists, and the analysis on a large number of domestic and international wareroom accidents, systematic filtering and classification are conducted on the risk types of the oil gas wareroom and the accident reasons. The principal reasons in the operating period of the underground oil gas storeroom for the layered salt rocks, are distinguished. The risk accident tree model during the oil gas storeroom is established. Through the minimum cut set of the risk accident tree model and the analysis of the structure importance degree, the main risk ranking during the oil gas wareroom operating period is acquired. The fuzzy comprehensive evaluation method is applied. Four different degrees, namely red, orange, yellow and green, for the main risk factor influencing extent during the oil gas wareroom operating period, are established. Furthermore, this provides effective guidance on the risk controlling of the wareroom operating.

6 Scientific exploiting system of the coal industry

6.1 New development on the backfill coal mining technology and the paste backfill mining technology

The backfill coal mining technology indicates that the backfilling materials, such as the gangues, the aeolian sands, the coal ashes, the construction wastes and the paste, are filled in the gob that is behind the working face, with the advancing of the coal mining working face. With the backfill coal mining technology applied, the rock mass movement and the ground subsidence can be controlled. This can release the significant transferring from the passive preventing and solving to the active preventing and solving for the mine site ecology and the safety production environment. It is the significant reforming of the coal production type. Currently, the paste backfill mining technology has been widely expanded and applied in China.

6.1.1 The rock strata movement controlling theory with the backfill coal mining [19]

6.1.1.1 The equivalent mining height theory for the rock strata movement analysis in the backfill coal mining

The equivalent mining height M_e: after the backfill is conducted, the ultimate roof subsidence. It is equivalent to exploiting anextremely thin coal seam, as shown in Figure 6-1.

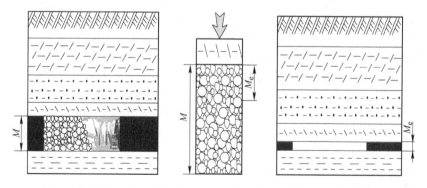

Figure 6-1　The schematic diagram showing the equivalent mining height

The expression equation of the equivalent mining height is:

$$M_e = (M - M_c) + (1 - K)M_c \tag{6-1}$$

Where, M_e is the equivalent mining height; M is the mining height; M_c is the roof subsidence; K is the compaction coefficient.

6.1.1.2 The rock strata movement and the ground subsidence calculation method with the backfill coal mining

The overlying rock strata movement is shown in Figure 6-2. Based on the equivalent mining height, the calculation expression of the overlying rock strata movement is:

$$w_z(x) = \frac{w_1(x)}{1 + \dfrac{H_1 - z}{H_0 - H_1 + L\tan\delta_0}} \quad (6-2)$$

Where, H_0 is the distance from the coal seam to the ground surface; H_1 is the distance from the fracture zone to the ground surface; z is the distance from the main core layer of the structure to the ground surface; L is half of the gob length; δ_0 is the rock strata movement angle.

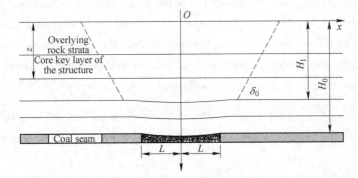

Figure 6-2 The schematic diagram of the overlying rock strata movement

Based on the equivalent mining height, the parameter determination method for the prediction of the ground surface subsidence:

    q—subsidence coefficient;

$\tan\beta$—the tangent of the main influencing angle;

    S—the offset distance of the inflection point;

    b—the horizontal movement coefficient;

    θ_0—the transmit angle of the mining influence.

6.1.2 The integrated technology with the backfill and coal miningcombined [19]

6.1.2.1 The integrated hydraulic support with the backfilling and coal mining combined

The fully mechanised solid backfill coal mining technology organically combines the fully mechanised coal mining and the fully mechanised solid backfilling. The simultaneous development of the coal mining and the backfilling in the working face is released. The arrangement for of the working face is basically equal to the arrangement form of the traditional fully mechanised working face. Meanwhile, the pattern of double working faces is formed. Specifically, the front working face is used to conduct the mechanised coal mining while the rear working face is used to conduct the mechanised backfilling. The core equipment in the fully mechanised solid backfill coal mining working face is the backfill coal mining hydraulic supports, as shown in Figure 6-3. Both coal mining and backfilling are conducted under the cover of the hydraulic supports.

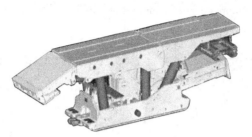

Figure 6-3 The integrated hydraulic support with
the backfilling and the coal mining combined

6.1.2.2 The designing flow path of the backfill coal mining under the buildings

Under the buildings, the designing flow path of the backfill coal mining is shown in Figure 6-4.

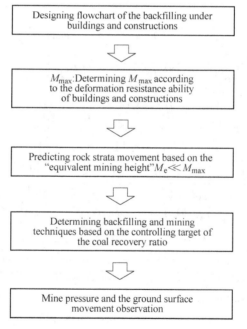

Figure 6-4 The designing flow path of the
backfill mining under the buildings

6.1.2.3 Engineering application and effects in the Mine 12# in Pingmei

For the Pingmeimine site 12#, the backfill coal mining technology is used to exploit the coal resources. For the mining and backfilling integrated hydraulic support, it is the positive four-bar linkage chocksupports with six columns, as shown in Figure 6-5. The maximum coal production quantity for one day is more than 3000 tons. The practial quantity of the gangues that are filled in the gob is more than 4800 tons. The average weight ratio between the mining and backfilling is 1.3. The maximum filling to mining ratio is up to 1.6, which is higher than the designing requirement. The backfill compactness is more than 92%. The integrated backfill and mining production technique is shown in Figure 6-6.

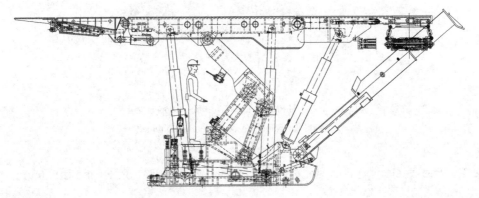

Figure 6-5 The schematic diagram showing the positive four-far linkage chocksupport with six columns

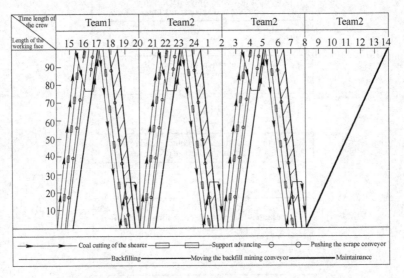

Figure 6-6 The mining and backfilling integrated production technique of the working face in the normal mining period

When the working face is in the mining period, the average working resistance of the hydraulic supports is ranged from 5500kN to 6000kN. The working resistance of the hydraulic supports has no abrupt ascending phenomenon. Additionally, for the working face, there is no periodic roof weighting phenomenon. The borehole drilling and imaging show that in the moving and deforming process of the rock strata, there is no apparent bed separation of the roof. There is no apparent collapsing zone. Furthermore, for the immediate roof of the overlying rock strata, there is no apparent rupture. Overall, the application effect of the backfill coal mining technology is favourable.

6.1.3 The technology and methods of recycling the strip coal pillar with paste backfill mining [20]

6.1.3.1 Fully mechanised coal mining method with full backfilling and large filling interval

On the ground surface, the gangue paste filling materials are prepared. Then, the backfill pump is used to apply the pressure. After that the pipe is used to conduct the transportation. The full gob area is filled with the material. The filling interval is ranged from 3m to 5m. Before each filling, along the working face, the isolation wall is prepared. The special new hydraulic support that has the isolation function and the supporting function for the area to be filled, is studied and developed. Also, the new combined isolation method that combines the flexible material that is matched with the hydraulic support and the plastic fabrics, is developed. The paste backfill requires the development of the special paste backfill isolation hydraulic support. The fully mechanised full filling coal mining method with the isolation and large filling interval, and the paste backfill hydraulic supports are shown in Figure 6-7 and Figure 6-8.

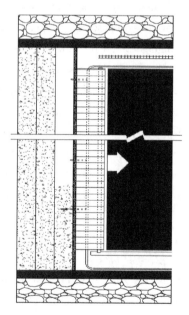

Figure 6-7 The fully mechanised full filling coal mining method with large filling interval

Figure 6-8 The paste backfill hydraulic support

6.1.3.2 Analysis of the application effect

The paste backfill technology to recycle the remaining coal pillars with the strip mining method is applied in the working face 2351 in the Daizhuang Coal Mine. For the paste backfill materials, they are fractured coal, fractured gangues, fine ashes, normal cement and the discharged water of the mine site. After the paste backfill reaches the full subsidence, the subsidence coefficient is 0.08. The paste backfill is an effective approach to recycle the remained coal pillars when the strip mining is used under the buildings. It can effectively protect the buildings and constructions on the ground. Meanwhile, it has the high exploiting ratio to recycle the remained resources when the traditional strip mining method is used. In the working face 2351 in the Daizhuang Coal Mine, the exploiting ratio in recycling the strip pillar is up to 91.6%. The roof contacting effect of the paste backfilling is apparent, as shown in Figure 6-9.

Figure 6-9 The roof contacting effect of the paste backfilling

6.2 The coal resource exploiting with the replacement of the waste gangues in the fold and fault area in which the high abutment pressure occurs[21]

The coal resource exploiting technology with using the waste gangues to replace the coal resources in the zone of folds and faults in which the high abutment pressure occurs, regards the high recycling ratio of the resources and the fact that the waste gangues do not need to be transported to the ground surface as the target. The mining system in which the waste gangues are used to replace the coal resources, is constructed. In the area in which the remained coal resources occur, the roadways with large cross-section area are excavated to recycle and exploit the coal resources. The waste gangues that are generated in the production process in the underground can be transported to the roadway with large cross-section area, based on the transporting system and the replacement system. This technology successfully solves the challenging problems such as the low recycling ratio of the coal resources under the complex condition and the processing of the waste gangues. It has pretty important significance in practice.

6.2.1 Geological production condition of the Xinsan Coal Mine

In the Xinsan Coal Mine, the waste gangue replacement exploiting zone is located in the syncline structural area. The wing angle of the coal seam 2# is degrees. The coal seam thickness ranges from 3.9m to 4.6m. The mean thickness is 4.5m. The coal mass is relatively fractured. The false roof is composed of the fine sandstones with a thickness of 0.34m and the carbon sandstone with the thickness of 0.3m. The immediate roof is the grey fine sandstone with the thickness of 2.3m. The main roof is the fine sandstone with the thickness of 2.5m. The immediate floor is the grey fine sandstone with the thickness of 3.18m. The challenges of the geological production conditions in the replacement exploiting zone, are mainly reflected in the following five aspects: the replacement exploiting zone is located in the axis of the syncline Linjiagou. The structural stress is high. The fractures of the coal seam roof is developed. The roof integrity is poor. The coal mass is fractured and the bearing capacity is low. For the two sides of the replacement exploiting zone, it is the overlapped zone of the gob abutment pressure. Three waste roadways such as the waste gangue replacement roadway and the Nanzheng roadway are intersected in space. The vertical distance ranges from 2m to 7m. In the replacement exploiting zone, there are 7 small faults with the vertical distance ranging from 1.0m to 2.5m. The layout of the working face waste gangue replacement roadways are shown in Figure 6-10.

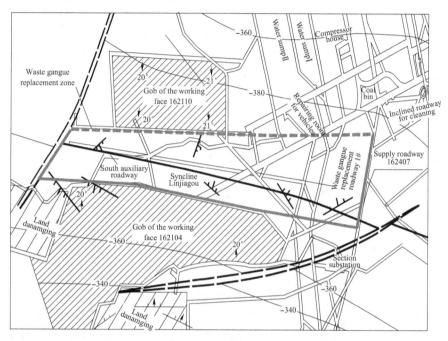

Figure 6-10 The layout diagram showing the working face waste gangue replacement roadways

Based on the above practial geological production condition, the characters of the surrounding rock masses of the waste gangue replacement roadway in the fold and fault zone in the Xinsan Coal Mine are studied, as following:

(1) The waste gangue replacement roadway with large cross-section is subjected to the influence of the structural stress of the Linjiagou syncline and the abutment pressure of the gob. The abutment stress is high.

(2) The coal masses and rock masses in the replacement zone are subjected to the violent geological movement such as the squeezing and torsion. The structural fractures of faults are developed. The coal masses and rock masses are fractured. The bearing capacity is low. The roof of the principal roadway along the south and the auxuliary roadway along the south in the replacement exploiting zone, is fractured. The bearing capacity of two sides is low. The two sides show outward moving integrally.

(3) Local area stress concentration occurs in the intersected area of the waste gangue replacement roadway and the above abandoned roadways. The roadway surrounding rock masses are fractured.

(4) The stress and the deformation of the surrounding rock masses increase dramatically in the square mode and cubic mode, wich is resulted by the large cross-section of the roadway. The fracturing range of the coal mass and rock masses that are around the coal roadway is large. Consequently, this makes that the anchorage force of the original rock bolts and cable bolts cannot be guaranteed. This largely increases the difficulty in the roadway supporting.

(5) In the traditional rock (cable) bolt reinforcement scheme design for the coal roadways, the roof cable bolts are installed in the single form. Consequently, when the cable bolts are installed, it cannot realise the pretension force along the horizontal direction. Furthermore, in the roof sinking process, the cable bolts that are installed in the single form cannot generate the reinforcement force along the horizontal direction. Particularly, it is not beneficial for the roadway roof to form the stable structure along the horizontal direction.

6.2.2 Support requirement and controlling scheme for the coal roadways with large cross-section in the high stress domain

6.2.2.1 The supporting requirement of the waste gangue replacement roadway in the Xinsan Coal Mine

(1) Improving the structure of the reinforcement system. In-situ practices show that the traditional rock (cable) bolt reinforcement cannot effectively solve the supporting challenges of the waste gangue replacement roadway in the Xinsan Coal Mine.

(2) The high strength rock bolt reinforcement is adopted. The high strength rock bolt reinforcement can provide relatively larger reinforcement resistance. It is required that the surrounding rock mass can have a certain deformation. The stress in the surrounding rock masses can be decreased. The rock bolt load can be decreased. Then, the rock bolt rupture can be prevented.

(3) The controlling of the core positions including the roadway side and the roadway corner, is strengthened.

(4) The stress concentration section in the intersected area between the waste gangue replacement roadway and the above roadway space, should be strengthened.

(5) For the fault fractured zone, the I-beam and the rock bolt (cable bolt) are coupled together.

6.2.2.2 The supporting scheme of the coal roadway with large cross-section in the domain with high stress

According to the deformation and failure characters of the surrounding rock masses, and the methods such as the numerical simulation calculation, the mechanical analysis and the engineering comparison, the whole scheme of the pretension cable truss joint controlling in the waste gangue replacement coal roadway with large cross-section in the high abutment pressure fold fault zone. The calculating results of the numerical simulation are shown in Figure 6-11.

In the high abutment pressure fold fault zone, under the normal condition, the newly developed high pretension cable truss (with the connector Ⅱ) is matched with the single pretesnion cable bolt, high strength and pretensioned rock bolt, steel abutting beam, and the metal mesh, to form the joint support to strengthen the roadway roof.

The joint support with the cable bolt-steel beam truss structure, the high strength threaded steel rock bolts, the steel abutting beam and the metal mesh, is used to strengthen two sides of the roadway. This makes the weak surrounding rock masses around the roadway surface form the intact integrity. This effctively control the deformation failure of the roadway. For part of

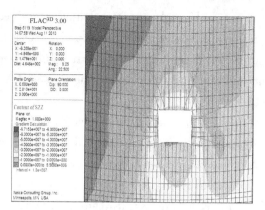

Figure 6-11   The vertical stress distribution diagram in the roadway surrounding rock masses

the roadway intersected zone and the fault fractured zone, the strengthening support measures should be conducted immediately. Under the normal condition, the specific supporting scheme of the fold and fault zone with the high abutment stress is shown in Figure 6-12.

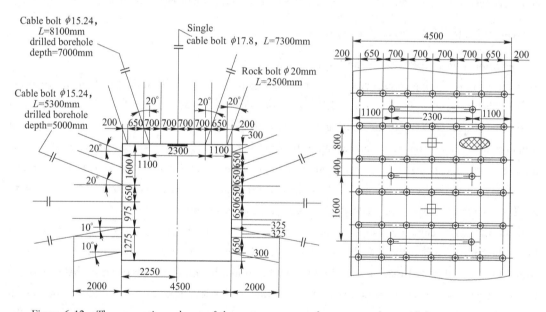

Figure 6-12   The supporting scheme of the waste gangue replacement roadway with large cross-section

6.2.3   The developing and techniques of the exploiting equipment for the coal resources with the waste gangue replacement

6.2.3.1   The developing of the waste gangue thrower equipment

A   The requirement on the function of the waste gangue thrower

(1) The waste gangue replacement roadway generates heaving due to the fact that the geological

structure in the replacement exploiting zone is complicated. To guarantee that the waste gangue can fully fill the area of the waste gangue replacement roadway, and it can adapt to the variation of the roadway height, the filling part of the waste gangue thrower should be able to realise the function of the adjustment along the up and down direction and the swinging along the right and left direction.

(2) The waste gangue thrower should be able to move backward automatically. When the waste gangues are fully filled in front of the waste gangue thrower, the waste gangue thrower should be able to move backward to a new working position to continue the filling.

(3) The transporting belt of the belt conveyor should be able to realise the automatic tension. The waste gangue thrower moves backward with a step. The transporting belt of the belt conveyor will loose. Therefore, it is required to conduct tensioning on the transport belt.

(4) To adapt the situation that the replacement roadway is uneven, it is necessary to design the automatic leveling device. Then, the automatic adjusting of the left and right level of the uninstalling level.

(5) When the walking chassis is moving along the retreat direction, it may deviate from the centre of the belt conveyor. Then, the automatic offset adjustment equipment is needed.

B The structure of the waste gangue thrower

a Determination of the form of the throwing part

To realise the designing requirement that the waste gangues should be fully filled in the roadway with the height ranging from 2.8m to 4.5m and the width of 4.5m, the kinds of the methods such as the belt transporting and the mechanical throwing can be used. Due to the fact that the belt transporting is an economic and reliable transporting method that is commonly used in the underground, and considering that the other methods have the problem on the aspect of flying dust and safety, it is finally determined that the belt transporting method is used for the throwing part.

b Determination of the form of the off tracking component structure

The off tracking component strcture is composed of the off tracking frame, the off tracking hydraulic lift, the off tracking valve and the anti-off tracking switch, as shown in Figure 6-13. Along the front side and rear side of the off tracking frame bottom, two supporting wheels and two horizontal wheels are set to support in the pathway of the levelling frame. One side of the off tracking lifting jack is connected to the off tracking frame and the other side is connected to the levelling frame. When the belt is off tracking, the belt will compress the anti-off tracking switch. The anti-off tracking switch controls the electromagnetic valve, which controls the swithing and shrinking of the off tracking lifting jack. Then, the off tracking action of the off tracking frame can be realised.

For the waste gangue thrower structure, besides the above mentioned throwing part and the off tracking component, the core structures such as the walking basement, the whole bottom basement, the leveling structure and the uninstalling roller are also included. The specific components and position are shown in Figure 6-14.

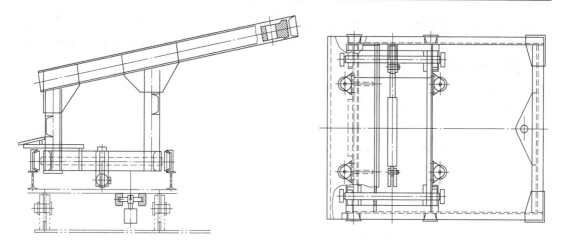

Figure 6-13 The structual diagram of the off tracking component

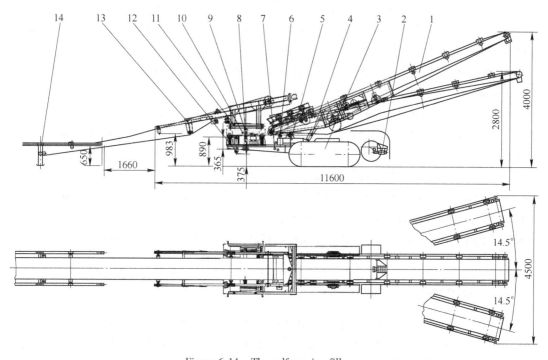

Figure 6-14 The self-moving fill part

1—Throwing betl; 2—Left and right braking lifting jack; 3—Moving basement; 4—Height adjusting lifting jack;
5—Rotation work bench; 6—Rotation lifting jack; 7—Whole bottom basement; 8—Leveling hydraulic lifting jack;
9—Off tracking lifting jack; 10—Leveling component; 11—Off tracking frame; 12—Supporting lifting jack;
13—Uninstalling roller group; 14—Compressing and rolling H frame

6.2.3.2 The backfilling technique to replace the coal resources with the waste gangue

When the waste gangues are transported to the head-on position of the waste gangue replacement roadway, the waste gangue thrower accomplishes the following tasks in sequence.

(1) The waste gangues are uninstalled to the belt that is used for conducting the throwing and

filling tasks from the belt conveyor end. In this full process, the waste gangues should not be spilled. When the direction of the belt conveyor is changing and the adjusting is conducted on the left side, right side, top section and bottom section of the belt that is used to conduct the throwing and filling tasks, the situation of the material feeding position changes. However, this has no influence on the material feeding task.

(2) The throwing and filling belt is used to throw and fill gangues to the head-on section of the roadway. When the belt that is used to conduct the throwing and filling task is working, the left side, the right side, the top section and the bottom section of the belt can be adjusted to make sure that the belt can be stretched to two sides of the roadway and the roof. In this case, the optimal backfilling effect can be realised.

(3) After the head-on roadways are fully filled, the waste gangue thrower should retreat for approximately 500mm. Following this, the belt frame is repealed and the belt is tensioned. The belt conveyor itself has the tensioning storage equipment. Therefore, this can be realised by pressing the buttons that are in the head of the equipment and in the end of the equipment to control the switch of the tensioning winch.

(4) After retreating for 2m, in the state that the belt conveyor is halted, the top carrier roller that is most adjacent to the backfilling section in the belt conveyor is removed. After that, the compressing and rolling H frame which is located in the middle of two longitudinal beams is retracted for approximately 2m. Following this, the machine is switched on and the backfilling is conducted again.

(5) After retracting for 4m, or when the backfilling section will touch the standard H frame of the belt conveyor, one part of the longitudinal beam and a standard H frame are removed. The compressing and rolling H frame is used to support the middle section of the standard H frame that is used for the next section. The self-moving backfilling section re-enters the next working cycle.

6.2.3.3 The mechanical model of the waste gangue throwing movement and its core technical parameters

The construction step of the waste gangue thrower directly influences the backfill quality and the construction velocity. In the horizontal roadway, the gangues that are thrown by the waste gangue thrower are accumulated together, similar to the cone. The angle of the cone is the natural repose angle of the rocks. Additionally, the height of the cone is the maximum height of the gangues in ascending. After the waste gangue thrower is retracted, the second cone is formed. Part of those two cones overlaps. In the top section of the waste gangue cone, undulating fluctuation occurs. The undulating fluctuation leads to the decreasing of the roadway backfill ratio. The mechanical model of the waste gangue throwing movement, which describes the relationship between the undulation amplitude and the construction step, is shown in Figure 6-15.

The bottom pedestal height of the waste gangue thrower is h. The belt that is used to conduct the throwing is MN. The waste gangues are thrown from the point N. After the peak point Q, the waste gangues arrive at the point D. The moving track is parabola. The initial velocity v and the angle α are the belt velocity and the angle of the belt in the waste gangue thrower that is used to conduct the throwing. ΔL is the distance between the waste gangue cone and the bottom pedestal of the waste gangue thrower before the waste gangue thrower retracts. $Q''E$ is one side of the waste gangue cone. As for the inclination angle, it is the natural repose angle of the waste gangues (θ).

The relationship equation between the area in which the backfill is not conducted and the backfilling step is depicted with the following equation:

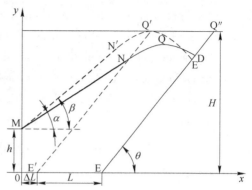

Figure 6-15 The mechanical model showing the waste gangue throwing movement

$$S = -\frac{1}{2}AL^2 + \frac{1}{2}(H - B - A\Delta L - AH\cot\theta + AC)L + \frac{1}{2}(H - B)(\Delta L + H\cot\theta - C) \qquad (6-3)$$

In this equation:

$$A = \frac{\cot\alpha}{1 - \cot\alpha\cot\theta}$$

$$B = \frac{\Delta L\cot\alpha + h + \frac{v^2}{2g}(\sin^2\alpha - 2\cos^2\alpha)}{1 - \cot\alpha\cot\theta}$$

$$C = \frac{H - h - \frac{v^2}{2g}(\sin^2\alpha - 2\cos^2\alpha)}{\cot\alpha}$$

The waste gangue throwing mechanical model indicates that the smaller the filling step, the higher the roadway filling ratio. In the domain of 0~0.5m, the variation of the filling step has no apparent influence on the roadway filling ratio. However, when the filling step is larger than 0.5m, the larger the filling step, the smaller the roadway filling ratio. When the filling step is 0.5m, the roadway filling ratio can be up to 99.3%. Considering that after the waste gangue thrower retreats for 4 times (2m), the carrying roller on a belt conveyor is removed. After retreating for 4m, the Hi frame on a belt conveyor is removed. It is finally determined that the filling step is 0.5m.

The Xinsan Coal Mine studied and developed the cable truss composite reinforcement technology for the waste gangue replacement roadway in which the construction techniques, the technical standard, the construction safety and matters needing attention, and the in-situ ground pressure monitoring are integrated. The experimental results indicate that in the test section, the rock bolt construction quality meets the experiment requirement. The convergence of the roadway cross-section is extremely small. The maximum convergence between two sides of the roadway is less than 145mm. The maximum roof convergence is less than 166mm. The controlling effect of the roadway surrounding rock masses is favourable, meeting the production requirement of the field.

In the filling process, the waste gangue thrower can swing along the upward, downward, right and left direction. Then, the levelling and the off tracking can be realised. The belt on the thrower can realise the function of self-pretension. The filling range of the waste gangue thrower is that the height ranges from 2.5m to 4.5m, and the width ranges from 3m to 5.2m. The quantity of the thrown waste gangues is 300t/h. The maximum throwing angle is 16 degrees. The throwning velocity is 3m/s. The filling step is 0.5m and the filling ratio reaches 99.3%.

6.3 The combined mining technology of coal and gases, and the underground gasification

Gas is the main restraining factor in realising of the safe and efficient exploiting of a number of coal mines. In China, for the majority of mine sites, the reserve of the gases in coal seams has the character of "two high with three low". Emphasis should be given to the pressure relieved gases that are influenced by the drainage mining activities. The underground gasification is beneficial for improving the recovery ratio and fully using the scrapped resources. Therefore, the multiple kinds of supply regarding the gasification, electricity generation, the heat supply and chemical products, can be realised. This provides development prospective for the development and utilisation of the hydrogen energy. Therefore, it fits the sustainable development strategy of China.

6.3.1 Combined exploiting of coal and gases [22]

6.3.1.1 Generation and dynamic variation of the elliptical paraboloid zone of the mining induced fractures

A Generation of the elliptical parabolid zone of the mining induced fractures

After the working face advances for a certain distance, the overlying rock strata that have the layered movement rupture which develops upwards successively and the bed separation character, will form the mining induced fracture zone of the overlying rock strata. Then, in the mining induced fractures of the overlying rock strata, the bed separation fractures are connected with the rupture fractures. In the space, the external boundary that is similar to the elliptical paraboloid surface, is formed. This is called as the external elliptical paraboloid surface. After the working face advances for a certain distance, the mining induced fractures of the overlying rock strata that is located in the centre of the gob, is basically compressed tightly. Its boundary can also be described with the elliptical paraboloid surface. Therefore, between the internal and external elliptical paraboloid surface, the mining induced zones that is similar to the cap, is formed. It is named as the elliptical paraboloid zone, as shown in Figure 6-16.

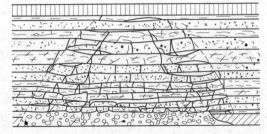

Figure 6-16 The plane distribution state of the mining induced fracture elliptical paraboloid zone

B The dynamic evalution law of the mining induced fracture elliptical paraboloid zone

With the working face advancing, the collapsing of the overlying rock strata and the bed separation height are influenced by the distance between the core strata and the exploited coal seam, and its strcutural stability state. This also indicates that the mining induced fracture eliptic parabolid zone is not a constant value. In fact, it is dynamically changed. With the working face advancing, first, before the 1st inferior core strata ruptures, the elliptical paraboloid zone forms. When the 1st inferior core strata ruptures, the middle of the elliptical paraboloid zone is compressed tightly. As for the surrounding area, the elliptical circle is distributed. Meanwhile, under the 2nd inferior core strata, the elliptical paraboloid zone is formed again. After the 2nd inferior core rock strata ruptures, the middle of the elliptical paraboloid zone is compressed tightly again. Also, for the surrounding area, the elliptical cycle is distributed. This continuously develops until the core rock strata. After the core rock strata ruptures, the elliptical paraboloid zone of the overlying rock strata does not occur. However, the fracture zone that the layered surface is distributed, is still a elliptical zone.

The zone width along the strike direction, has a closed relationship with the initial weighting step and the periodic weighting step when the working face is being exploited. Around the open-off cut, the width of the elliptical paraboloid zone is approximately equal to 1 times of the initial weighting step. Around the working face, the width changes between 2 times of the periodic weighting step and 3 times of the periodic weighting step.

The width along the dip direction also has a closed relationship with the initial weighting step when the working face is being exploited. It is approximately 0.7 ~ 0.8 times of the initial weighting step. The width along the dip direction is determined by the surrounding abutment condition. For example, when the coal seam is almost horizontal or the gently inclined, and the abutment condition is equal, the widths around the intake air roadway and the return air roadway are equal.

6.3.1.2 The theory of the drainage of the pressure relieved gas in the mining induced fracture elliptical paraboloid zone

A The diffusion process of the pressure relieved gases

For the stability state of the stable air which has the gases included, it means the balance is aquired between the pure diffusion and the pressure diffusion. The gravity of the air generates the pressure gradient which is along the downward direction. However, the direction of the diffusion flow which is generated by it, is reverse with the pressure gardient. The gases have the tendency to diffuse upwards. It means that with the increasing of the altitude, the gas concentration increases accordingly.

B The principle of the pressure relieved gas drainage in the mining induced fracture elliptical paraboloid zone

Under the rising and floating diffusion, and the seepage power effect, the pressure relieved gases move to the zone where the fractures are fully developed, along the fracture network of the mining induced fracture zone. Therefore, the movement and reserve zone of the pressure relieved gases, is mainly the top fracture zone and the surrounding fracture development circle, in the first stage of the mining induced fracture development. In the second stage, they are mainly concentrated in the surrounding fracture development circle. The pressure relieved gas drainage principle is shown in Figure 6-17.

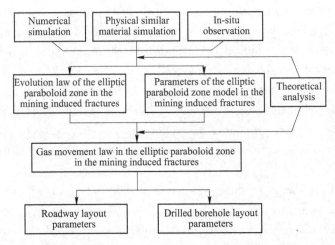

Figure 6-17　The schematic diagram showing the pressure relieved gas drainage theory

The mining induced fracture ellipitical paraboloid zone is the place where the gases are concentrated and flowed. If the end of the drainage hole (or the end of the drainage roadway) is set in it, it is beneficial for the pressure relieved gases to leak to the drainage hole (or the drainage roadway). Therefore, the drainage time of the drainage hole (roadawy), the area and the effect, can be guaranteed. The layout of the drainage roadway is shown in Figure 6-18.

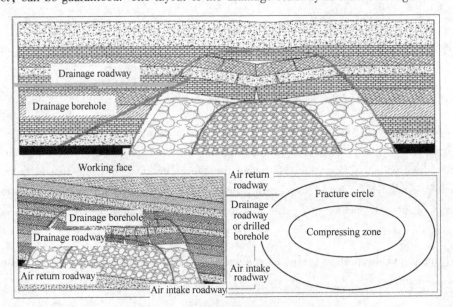

Figure 6-18　The arrangement diagram along the strike direction when the initial weighting comes

6.3.1.3 Practices of the combined exploiting of the coal and gases

A The pressure relieved gas drainage technology with the high drainage roadway

The main method to drain gases in the working face is that using the high drainage roadway along the strike direction in the roof (which is abbreviated as the high drainage roadway) is used to drain the gases that are in the adjacent coal and rock masses, and the gob. The end of the roadway is set in the mining induced fracture ellipitical paraboloid zone, as shown in Figure 6-19 and Figure 6-20.

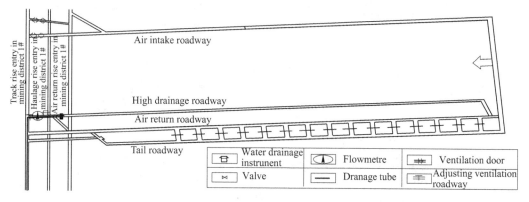

Figure 6-19 The working face roadway arrangement diagram

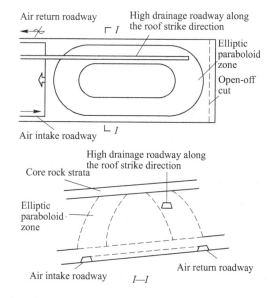

Figure 6-20 The schematic diagram showing the high drainage roadway arrangement

B The relationship between the drained gas concentration in the high drainage roadway and the layer position

Under the normal production period, the mean gas drainage ratio in the high drainage roadway is up to 58%. The gas concentration of the tail roadway (0.46% ~ 2.36%) is basically controlled under 2.5%. As for the working face (0.12% ~ 0.76%), the top roadway corner (0.09% ~ 0.77%) and the return air roadway (0.3% ~ 0.94%), the gas concentration are controlled within in 1%. Therefore, the safe and efficient production of the working face is realised. The relationship between the drainage gas concentration in the high drainage roadway and the layer position is shown in Figure 6-21.

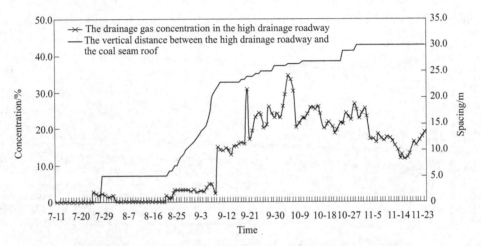

Figure 6-21 The relationship between the gas concentration in the high drainage roadway and the layer position

6.3.2 The underground gasification

6.3.2.1 New techniques of the underground gasification

Long roadway and large cross-section. The character of the underground gasification in two stages:

(1) The gasification mode in which the shaft included and the shaft excluded are combined.

(2) For the furnace building engineering, the coal roadway is regarded as the main part. It is easy to operate and expanded.

(3) Two coal roadways which are set along the coal seam inclination direction ($> 4m^2$), are regarded as the air-out tunnel.

(4) The air supply to the underground is conducted following two stages. In each stage, the air, oxygen, water vapour with different percentage, are used to supply the air.

6.3.2.2 The establishment of the underground gas producer and ignition

The ignition method of the gas generator with the shaft excluded:

(1) The penetration method of the air power: In the middle, the hole is drilled. And the blast connection method.

(2) The electric penetration method.

(3) New method: The atomic energy blasting method; In 1967, the USA used the chemical liquid blasting method.

6.3.2.3 The technical system of the underground gasification production

A Exploiting of the gasification area and the selection of the gasification type

The exploiting along the strike direction: (1) from the middle to the boundary; (2) from the boundary to the middle; (3) from the boundary to the boundary; (4) from one side to the boundary and from the middle to the other side.

The exploiting along the dip direction: (1) the continuous exploiting from the bottom to the top; (2) exploiting with different stages, from the bottom to the top; (3) exploiting with different stages, from the top to the bottom (when the dip angle is large).

Attention should be paid that it should be matched with the position of the ground surface gasification station. The mininum air supply path should be selected.

B Main outcomes of the underground gasification

(1) More than ten million gases for motor fuel have been produced effectively for a long term. The underground gasification technology for the coal seam that is buried below 350m has been successfully grasped. In the soft coal seam and brown coal seam, the method and the flow path by which the industrial feed gases are produced directly from the underground gasification, are formulated. The kinds of methods to develop the gasification channels in the coal seam, are grasped. Furthermore, the new cut-through method and the directed drilling cut-through method in the thin coal seam with a depth up to 500m, are studied and developed.

(2) The gas turbin fan with high efficiency and large volume, has been produced in batch. The volume is up to $2500 \sim 3260 m^3/min$. From the perspective of technology, the underground gasification issues of the lean coal and the meagre coal with the carbon content of 90%~92% and the volatile matter of 6%~9%, are basically solved.

(3) The gasification and unwatering method of the coal field is successfully grasped. Furthermore, the exploiting drilled boreholes and the unwatering equipment are correspondingly designed and produced. The management and controlling method of the gasification process, is established. The new gasification system is tested, such as the filling gasification system of the blind drilled boreholes.

(4) The instrument that is used to measure the ground subsidence, and the leading edge, the shape and the temperature of the flame working face, and the indicator to measure the gas leakage, are developed. This provides condition for the controlling and management of the gasification process. For the exploiting of the inferior coal seam, the thin coal seam and the deep coal seam, it is quite difficult to acquire the satisfactory technical benefits and results. However, the underground gasification is an effective technology to solve this issue.

6.4 The exploiting of the mine without drainage and the underground water storage technology [23]

In China, the loss of the mine water is massive. The mine water loss is the important technical issue that is confronted by the coal green exploiting. The west of China is the main coal production zone in which the production quantity is maximal in the world while the water shortage is severest. The mine exploiting without drainage and the underground water storage technology, effectively realises the underground long-term safe and economic storage of the mine water. This greatly increases the mine water utilisation rate in coal mines that are in the west of China.

6.4.1 The underground reservoir construction technology of coal mines

6.4.1.1 Evaluation method for the selection of the underground water storage place

A Principle of the place selection

(1) The coal seam floor should not leak water. The floor such as the mudstone which has the favourable water proof and with the faults excluded, and the gob in which the stability of the coal pillar is favourable, are used to reserve water.

(2) The gob area can be used to accumulate water. According to the revealed underground water field of the mine site, the area that is beneficial for the accumulating of the mine water, is used to store water.

(3) Proper exploiting programming and water transfer. The exploiting programming and the water drainage system are optimised. The area which is beneficial for the mine water importing and exporting, is selected to store water.

B The comprehensive evaluation results

The comprehensive evaluation results indicate that in the Shendong mine site, more than 60% of the gob is applicable for storing water. However, currently, the practial water storage area is less than 20% of the gob area. This demonstrates that the coal mine gobs in the west of China, has enough storage area to store mine water. The evaluation of the water storage address is shown in Figure 6-22.

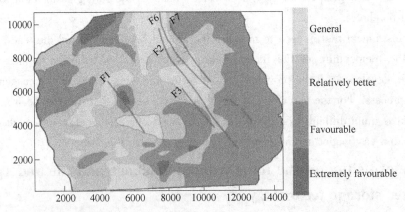

Figure 6-22 The evaluation of the water storage address in the Daliuta Mine

6.4.1.2 The underground reservoir volume expanding technology

First, the mining technology for the extremely large working face is developed. The problems such as the high ground pressure and the difficult ventilation, are solved. The largest working face in the world is established. This makes the water storage ability of a single gob increase 2 to 3 times. Furthermore, several gob areas are combined. This makes those several gobs combine together, becoming the underground coal mine reservoir which has the large water storage ability. This makes the water storage ability increase 10 times than the single gob area. The underground revervoir volume expanding technology is shown in Figure 6-23.

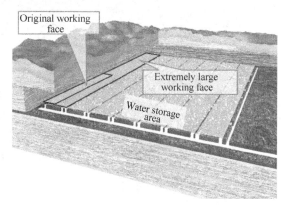

Figure 6-23 The schematic diagram showing the underground reservoir volume expanding technology

6.4.1.3 The designing method of the underground reservoir dam

The coal mine underground reservoir structure is particular. It is composed of coal pillars and the artificial dam. It is inharmonous, discontinuous and cross-section changing structure. Meanwhile, the load suffering is complicated. Specifcially, it is subjected to the composite effect of mining activities, the ground pressure, the water pressure, particularly the mine earthquake and the earthquake.

A Influence law of the earth on the underground reservoir dam

The numerical simulation and the physical simulation are conducted on the stress and deformation of the dam under the earthquake condition, as shown in Figure 6-24. The simulation results are shown in Figure 6-25. The results show that under the earthquake conditon, the safety of the underground reservoir dam in the coal mines is much higher than the ground reservoir. For example, when the earthquake intensity is 10, the safety degree of the underground reservoir dam in the Daliuta Coal Mine is 3.83 times of the ground reservoir.

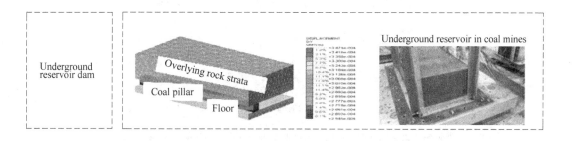

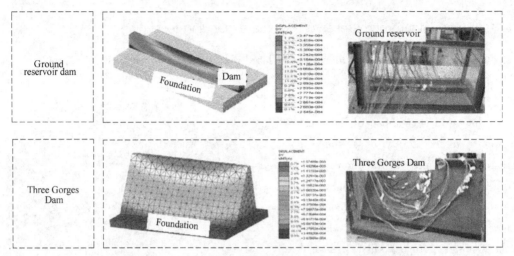

Figure 6-24　The numerical simulation and physical simulation

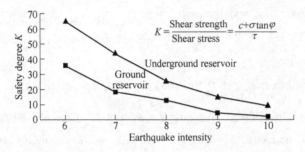

Figure 6-25　The physical simulation results of the safety degree of the different reservoir dams

B　The calculation methods for the two main factors of the underground reservoir dam

Based on the numerical analysis and calculation, the calculating methods of two main parameters factors of the underground reservoir dams under the influence of the mining activities, the ground pressure, the water pressure and the earthquake, are proposed. It is acquired that in the Shendong Mine Site, the width of the coal pillar dams ranges from 20m to 30m. The thickness of the artificial dams is around 1m. The calculating model of the width of the coal pillar dam and the thickness of the artificial dams is shown in Figure 6-26 and Figure 6-27.

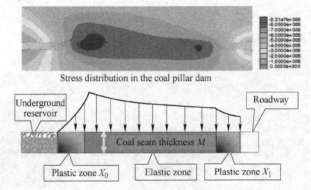

Figure 6-26　The calculating model of the width of the coal pillar dam

6.4 The exploiting of the mine without drainage and the underground water storage technology

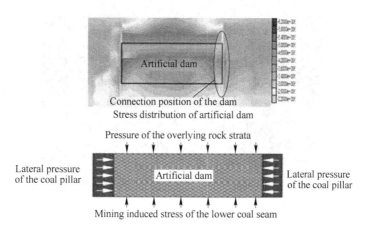

Figure 6-27 The calculating model of the artificial dam thickness

The width of the coal pillar dam:
$$Y = X_0 + K_0 M + X_1 \qquad (6-4)$$
Where, $X_0 \setminus X_1$ are the width of the plastic zone; K_0 is the adjustment coefficient; M is the coal seam thickness.

The thickness of the artificial dam:
$$S = \frac{K_1 PF}{\tau L} \qquad (6-5)$$
Where, K_1 is the adjustment coefficient; P is the water pressure resistance strength; F is the cross-section area of the dam; τ is the shear strength of the dam; L is the perimeter of the artificial dam.

6.4.1.4 The connecting method of the artifical dam and the coal pillar dam

(1) The joint between the artifical dam and the coal pillar dam is the weakest position of the underground reservoir structure strength. To guarantee the safety of the dam structure, the cutting connection method between the artifical dam and the coal pillar dam is invented, as shown in Figure 6-28.

(2) The calculating model for the cutting depth is proposed. The specially used cutting equipment is invented, as shown in Figure 6-29. It is studied that for the Shendong Mine Site, the cutting depth is 0.3~0.5m.

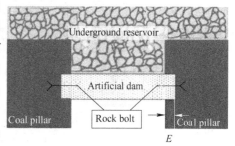

Figure 6-28 The schematic diagram showing the cutting connecting technique

$$E = \frac{K_2 PF}{\delta L} \qquad (6-6)$$

Where, K_2 is the adjustment coefficient; P is the water pressure resistance strength; F is the cross-section areaw of the dam; L is the perimeter of the artificial dam; δ is the smaller value between the compressive strength of the artificial dam and the coal pillar dam.

Figure 6-29 The cutting equipment of the artifical dam in the underground reservoir

6.4.2 The safe operation technology of the underground reservoir in coal mines

6.4.2.1 The connection technology of the upper and lower layer reservoir

(1) It reveals the influence on the lower layer coal seam exploiting on the safety of the reservoir in the upper layer coal seam. The safety distance along the horizontal direction between the exploiting position of the lower layer coal seam and the undergrond reservoir in the upper coal seam, is determined. The influence of the mining induced stress distribution with the exploiting of the lower coal seam, and its influence on the stress of the upper layer rervervoir dam, are shown in Figure 6-30 and Figure 6-31.

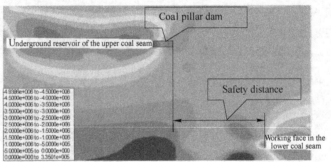

Figure 6-30 The mining induced stress distribution in the lower layer coal seam

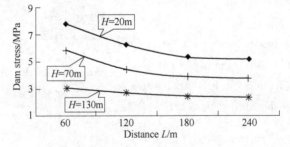

Figure 6-31 The influence of the exploiting of the lower layer coal seam in the Daliuta
Mine on the stress of the upper layer reservoir dam

H—Distance between coal seams; L—Distance between

the working face in the lower coal seam and the dam

(2) The shaft drilling technology with large calibre for the underground mine is developed. In the Daliuta Coal Mine, the upper layer reservoir and the lower layer reservoir with the vertical distance of 155m is successfully connected. The water transfer channel with the diameter of 1.4m is constructed. There are four channels for water transport and distribution. Among them, for two channels, the diameter is 423mm. As for the rest, they are 219mm and 159mm. Between the upper coal seam and the lower coal seam, the water transferring channel of the reservoirs, is shown in Figure 6-32.

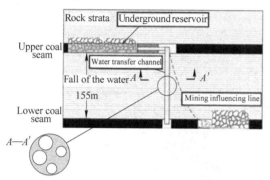

Figure 6-32 The water transferring channel of the reservoirs between the upper coal seam and the lower coal seam

6.4.2.2 The deploying method of the water between the underground reservoirs

(1) The deploying between the reservoirs in the same layer: When the water level of a certain reservoir is relatively higher, it is transferred to the reservoir in the same coal layer. The safety of the reservoir in the same layer is guaranteed, as shown in Figure 6-33.

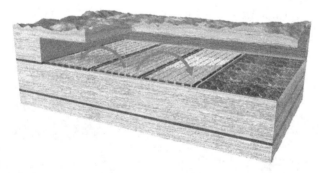

Figure 6-33 Transferring the water into the reservoir in the same coal layer

(2) The deploying to the reservoir in the upper layer: When exploiting the lower coal seam, the mine water is pumped and drained to the reservoir in the upper coal seam. Meanwhile, the reservoir in the lower coal seam is established, as shown in Figure 6-34.

(3) The deploying to the reservoir in the lower layer: Before exploiting the coal seam that is below the upper coal seam reservoir, the water in the reservoir is transferred to the lower coal seam reservoir. And then, the coal seam is exploited, as shown in Figure 6-35.

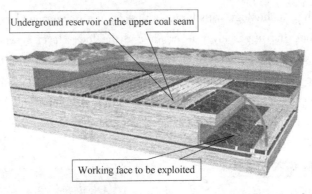

Figure 6-34　Transferring the water to the reservoir in the upper coal seam

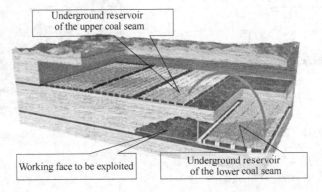

Figure 6-35　Transferring the water to the reservoir in the lower coal seam

6.4.2.3　Triple preventing and controlling technology for the reservoir operation safety

(1) The safety monitoring method for the underground reservoir safety is invented. The real time monitoring system for the parameters such as the mine earthquake, the earth, the water level and the dam, is developed, as shown in Figure 6-36.

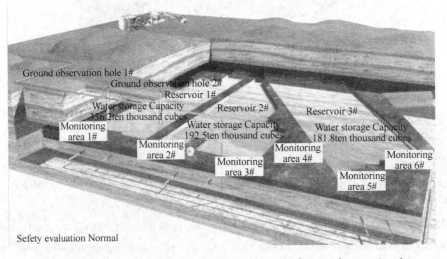

Figure 6-36　The monitoring system for the coal mine underground reservoir safety

6.4 The exploiting of the mine without drainage and the underground water storage technology

(2) The triple preventing and controlling technology for the reservoir safety operation (as shown in Figure 6-37).

I The water transferring between reservoirs to guarantee the whole safety of the reservoirs.

II The stress and deformation monitoring of the dam to guarantee the dam safety.

III The emergency drainage to gurantee the reservoir safety under the particular condition.

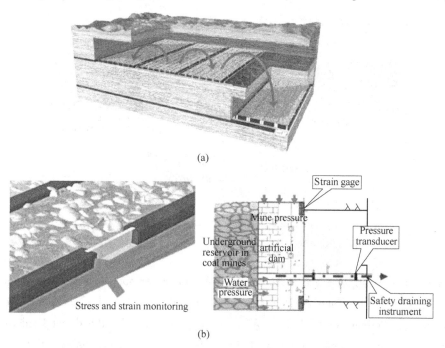

Figure 6-37 The triple preventing and controlling technology of the underground reservoir in coal mines
(a) Water transferring technology between reservoirs;
(b) Dam stress and strain monitoring technology emergency drainage technology of artificial dam

6.4.2.4 The water quality guaranteeing technology

The trinity water quality controlling technology of the underground reservoirs in coal mines is developed. This is to assue the safety of the water quality in the reservoir and the water quality of the underground water in the zone, as shown in Figure 6-38.

(1) Pre-processing before entering the chamber: The pre-processing such as the sediment and filtration, is conducted on the mine water that is slightly polluted, before it enters the chamber.

(2) Purifying of the water quantliy in the reservior: The rock masses that are collapsed in the gob area, are fully used to conduct the purifying processing on the mine water.

(3) Special processing in the underground: Special processing is conducted on the mine water that is polluted severely in the coal mass exploiting in the underground.

6.4.3 Application effect

The underground water storage technology is fully applied in the Shendong Mine Site. More than

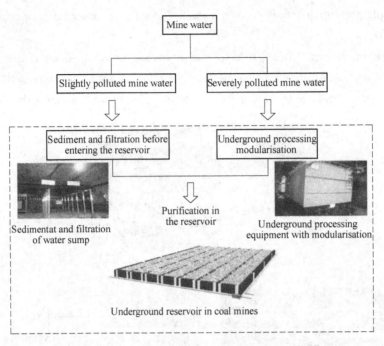

Figure 6-38   The trinity water qualtiy controlling technology

30 underground reservoirs are established. In 2014, the water supply is more than ten million cube. The majority of the water ultilisation in the mine site is supplied. Specifically, it supplies water for the surrounding power station. Also, it supplies water for the direct liquifying of the coal. This makes the coal mines in the west change from the large water consumption to the water supply basement. Meanwhile, it provides the water source for the ground ecology. The vegetation coverage in the mine site is relatively higher. The desertisation mine site in the paste changes to the oasis. This promotes the relatively better repairing of the ground ecological environment.

6.5 Intelligent mining technology [24, 25]

The intelligent mining is the new stage in the development of the fully mechanised mining technology of coal resources. It is also the compulsory requirement of the technology reforming of the coal industry and the upgrading development. In the "Zhong Guo Zhi Zao-Neng Yuan Zhuang Bei Shi Shi Fang An", the green and intelligent mining and excavating equipment for coal materials is listed as one of the development tasks of the energy equipment. In the 13th five-year plan of China, in the item of the core technical equipment of energy, it is explicitly proposed that boosting the research and application of the manless mining technology in coal masses should be accelerated.

The demonstrating engineering is located in the Tongxin Coal Mine Limited Company belonging to the Datong Coal Mine Group. The mean thickness of the coal seam in the fully mechanised caving working face 8202 is 15.26m. The mean dip angle of the coal seam is 1.5 degrees. The coal seam is relatively stable. The immediate roof is the carbon mudstone with a thickness of 3.87m. The main roof is the sandy mudstone with a thickness of 17.04m. The immediate floor is the sandy mudstone with a thickness of 1.98m. The length of the working face is 200m and the advancing length is 1500m. The production ability of the fully mechanised working face is 3000t/h. Then, in this demonstrating engineering, the total production for one year arrives at 10 million tons. The advanced technical equipment, the mature coal caving production technique, the reliable safety preventive measures, management method and the system are formed. Then the exploiting and mining of the fully mechanised caving working face with ten million tons is accomplished. It reaches the advanced technology and the economic index.

6.5.1 The electro-hydraulic remote coordinated controlling system with groups

In the fully mechanised caving working face, the electro-hydraulic remote coordinated automatic controlling system with groups, systematically combines the electronic controlling system of the coal shearer, the electro-hydraulic controlling system of the hydraulic supports, the communication controlling system for three machines in the working face, the pump station controlling system, the communication controlling system of the belt conveyor in the horizontal roadway and the power supply system. Then, they are connected into the automatic system of the mine. This realises the functions of the concentrated controlling, as shown in Figure 6-39. This system realises the remote concentrated controlling of the equipment in the working face, the interlocking of the equipment controlling system, the working status of the equipment in the working face with the remote observation, the computer management networking of the production information in the working face.

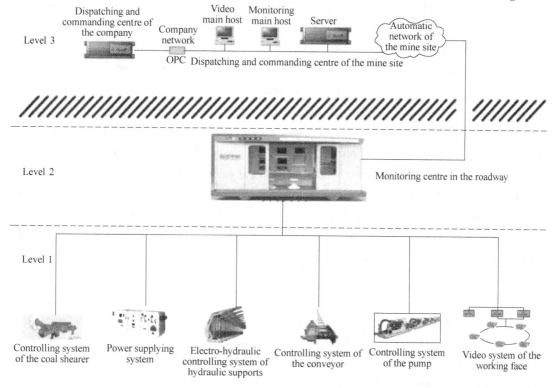

Figure 6-39 The structural diagram showing the automatic system of the fully mechanised caving working face

In the Tongxin Coal Mine, the electrohydraulic controlling of the four-post chock-shield-type coal caving hydraulic supports, is coordinately controlled by four vertical columns. Also, the support posture is controlled by the dip angle transducer. The coordinate controlling is conducted between the coal cutting, the moving of the hydraulic supports and the coal caving. The reaction speed of the automatic fluid infusion guarantees the support quality of the hydraulic supports. This realises the real-time online observing of the working situation of the main production equipment. The hidden malfunction can be revealed in time to avoid failure of the equipment. The operating rate of the quipment is improved. Then, the automation of the production process in the fully mechanised working face is realised, to reduce the labour strength and improve the production efficiency. The hydraulic supports used in the Tongxin Coal Mine is shown in Figure 6-40.

Figure 6-40　The hydraulic support ZF15000/27.5/42D

6.5.2　Reliability of the set of the equipment system in the fully mechanised mining

The production system of the fully mechanised caving working face is a complicated system which is composed of multiple procedures, multiple links and multiple equipment. As a consequence, in the production, the reliability of each production equipment will directly influence whether the reliability of the whole production system is high or low. Additionally, it will influence whether the production is large or small. Moreover, it will influence whether the economic benefits of the whole mine site are good or bad. For the automatic working face which has already realises the remote coordinated controlling with groups, the concentrated application of a large quantity of new technologies make the reliability of each equipment become more important in the production process of coal mines. As a consequence, improving the reliability of the equipment and system in the fully mechanised caving working face is the core parameter in successfully realising the remote coordinated controlling of the equipment with groups. The specific improving measures include the following aspects, such as the improving of the reliability of the single equipment, the application of the intelligent cleaning industry system, the remote diagnosing of the equipment in the working face, and maintaining of the equipment in the working face.

6.5.2.1　Intelligent cleaning and liquid feeding system

(1) The emulsion pump station is the core power of the hydraulic supports for the whole working face. To guarantee the requirement of the large flow, high reliability and high velocity of the hydraulic supports in the working face, the emulsion pump station uses the filtering technology with multiple levels and the intelligent controlling technology. This provides the powerful guarantee for the production of the modern coal mines with high production, high efficiency and safety. The

filtering system with multiple levels, is shown in Figure 6-41.

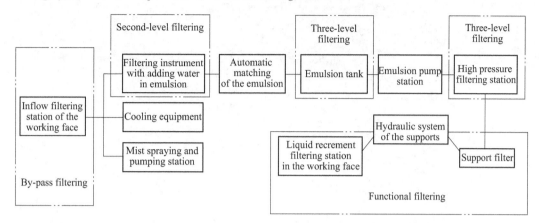

Figure 6-41 The filtering system with multiple levels

(2) The intelligent controlling technology of the liquid feeding system.

1) The liquid feeding system uses the intelligent distribution controlling.

The emulsion pump station uses the distribution controlling to improve the stability of the system. Meanwhile, among each controller, the highway can be used to conduct the interactive communication. It can control and manage intensively.

2) Application of the frequency conversion.

The frequency conversion technology has the functions of speed governing, energy conservation and soft starting. The joint ultilisation of the frequency conversion technology and the electromagnetic unloading valve of the pump staiton, makes the emulsion pump station supply more stable and effective hydraulic source for the working face.

3) Improving the interaction of the liquid feeding system.

The intelligent liquid feeding controlling system can conduct the automatic detection, real-time displaying and controlling for the whole system. Meanwhile, it has the functions of data recording, data saving and uploading. The operation information is transportted to the centralised controlling centre of the working face in time. Also, the sharing is realised through the computer network. And the informatisation of the production management is realised. The whole structure of the intelligent integrated liquid feeding is shown in Figure 6-42.

6.5.2.2 The remote diagnosing and maintaining of the working face equipment

According to the measured data, the coal shearer judges whether the equipment is in the malfunction state. And the malfunction type is distinguished. The system saves the data of all meausring equipment. Meanwhile, the historic data can be displayed through the screen at any moment. The computers on the ground can realise the monitoring of all coal shearer information and the adjustment of the parameters. Then, the remote malfunction diagnosing and maintaining of the coal shearer can be realised, as shown in Figure 6-43.

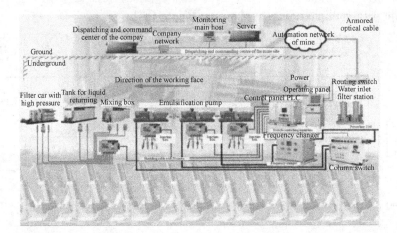

Figure 6-42 The whole structure of the intelligent integrated liquid feeding

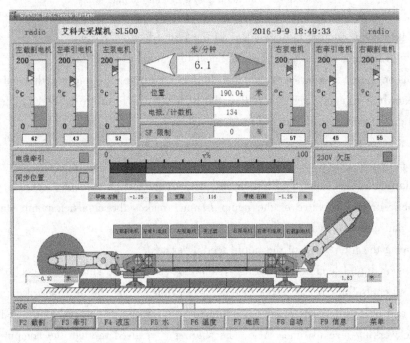

Figure 6-43 The remote diagnosing and maintaining of the coal shearer

6.5.3 Self-moving hydraulic supports following the shearer and the intelligent coal caving technology

6.5.3.1 The automatic coal caving of the fully mechanised caving working face

Compared with the normal working face, the coal mining techniques of the top coal caving working face is more complicated. To be more specific, it requires that the electric controlling system of the hydraulic supports has the complete controlling strategy on the coal caving process. Therefore, it is necessary to develop the controller and the controlling software for high-end hydraulic supports, improving the coal caving effect. The study on the automatic coal caving techniques is conducted. The multiple coal caving techniques such as the coal caving in a single round, the coal caving in multiple rounds are realised. Furthermore, based on the algorithm, comprehensive processing is conducted to acquire the analysis of the character signals of the multiple transducers. Furthermore, the information is reported back to the controller of the high-end hydraulic supports, realising the closed ring controlling of the coal caving ports. Then, the automation of the fully mechanised top coal caving process is realised.

For the coal caving hydraulic supports, the corresponding actions of the coal caving in the rear section include stretching of the rear beam, contracting of the rear beam, stretching of the plugboard, contracting of the plugboard, coal caving and mist spraying. The hydraulic support controller conducts controlling on the coal caving switch according to the corresponding time parameter values.

The semi-automatic top coal caving method or "the combination of the time controlling and the manual interfering", means that the operators initiate the groups and the automatic coal caving is controlled based on the time. Additionally, under the necessary condition, the semi-automatic coal caving is conducted with the operators finding the single hydraulic support. According to the loading of the rear scraper conveyor, the automatic controlling of the quantity in the coal caving port and the number of groups will be realised.

According to the reserving condition of the coal masses and rock masses in the working face, the coal mass and rock mass caving law under the supporting with the chock-shield-type hydraulic supports with four columns, the theory and the analysis based on the in-situ experimental observations, the appropriate parameters of the coal caving technique are determined. To be more specific, these include the influences of the coal caving step, the coal caving method and the variation of the reserving conditions of the coal seams. Furthermore, the mining and caving orders of the working face can be optimised. The automatic coal caving flowchart of a single hydraulic support is shown in Figure 6-44.

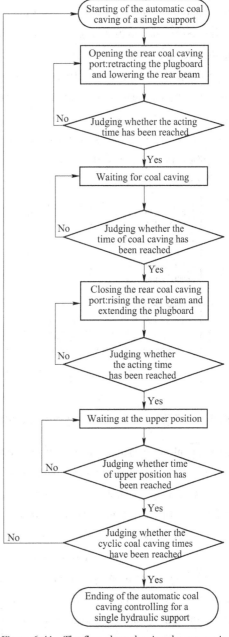

Figure 6-44 The flow chart showing the automatic coal caving of a single hydraulic support

6.5.3.2 The coal caving with memory of the hydraulic supports in the fully mechanised working face

The coal caving with memory indicates that the hydraulic support operating staff conduct a comprehensive top coal caving operation, according to the above mentioned automatic top coal caving process of a single hydraulic support, as shown in Figure 6-45. Based on the data analysis and the algorithm, the main host gathers the action information and generate a series of memory parameters, according to the method of the timer shaft. Then, through the highway, the data are delivered to the controller of the hydraulic supports in the working face. The controller of the hydraulic supports can conduct the coal caving process, according to the memory parameters. This reproduces the manual operating process of the hydraulic support operating staff with the maximum extent. The coal caving algorithm with memory can fuse and record the manual operation flowchart techniques. This makes the coal caving technical parameters of the fully mechanised working face more reasonable. It plays an important role in improving the coal caving efficiency and increasing the production of the working face.

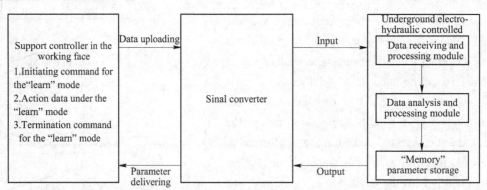

Figure 6-45 The schematic diagram showing the data flowchart of coal caving with memory

6.5.4 Digital mine management platform system

The digital mine management platform system is shown in Figure 6-46.

(1) The basement network platform: Through the bandwidth fibre ring network, the in-situ highway and the connected database, the interconnection of the information, the reliability, the safety and the timeliness are guaranteed.

(2) The mine database warehouse: According to the geology, the production information, and the normative database structure, the information island and the data redundancy are eliminated.

(3) The automation system of the mine industry: This including the monitoring system of the underground belt, the monitoring system of the mine site power supply, the monitoring system of the mine site drainage, and the monitoring system of the main fan room. Through the systematic integrated platform, the effective collaboration can be performed. Then, the automatic controlling and the remote controlling can be realised.

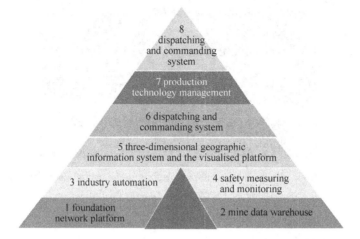

Figure 6-46 The digital mine management platform system

(4) The measuring and monitoring system of the safety production: The monitoring system of the safety production mainly includes the wireline communication, the wireless mobile communication system, positioning of the mine staff, the positioning system of the underground mobile equipment, the video monitoring system, the measuring system of the safety production environment, the monitoring system of the hydrology, the measuring system of the ground pressure, the monitoring system of the fire hazard, and the monitoring system of the ground surface movement. Through them, the actual time can be supplied for the data warehouse, the safety management and the scheduling system.

(5) The three-dimensional geographic information and the visualisation integrated platform: Based on the uniform three-dimensional space and time coordinate, all information in the mine site can be visualised. Then, the true three-dimensional mine site geographic information model can be established.

(6) The pre-controlling and managing system of the safety hazard: The transparent management of the hidden hazards for all objects under the ground surface is realised. This is to realised the mine hazard pre-controlling and pre-managing object that the labours, the machines, the environment and the management are highly syncretic.

(7) The production technical management system: This includes the geographic information system, the synergetic designing system of the mining engineering, the intelligent ventilation and prevention system, the geographic information system of the power transmission and distribution, the geographic information system of the industrial network and the management system of the electromechanical device.

(8) The dispatching and controlling system: The safe measuring and monitoring is realised. Also, the management integration of the safety hazard is realised. The system has the decision-making analysis, the malfunction diagnosing function, and the rapid reaction ability in the disaster period. The remote monitoring platform of the control centre is shown in Figure 6-47.

Figure 6-47 The remote monitoring platform in the control centre for the working face 8202 in the Tongxin Coal Mine

6.5.5 Application effect

The intelligent exploiting technology is applied in the working face 8202 in the Tongxin Coal Mine. In the working face, the accumulated production of the raw coal for three months is 3155.67 ten thousand tons. The recovery ratio of the working face is 86.41%. The comprehensive information utilisation rate of the fully mechanised equipment, such as the hydraulic supports in the working face in the Tongxin Coal Mine, is improved. The underground working condition is improved. The number of roof accidents is decreased. This promotes the safe, highly efficient, and integrated development of the coal mine industry. Also, the exploiting mode of safety, high efficiency, intelligence and green, in which one mine with one working face in the fully mechanised caving working face with ten million tons, is realised. This provides excellent demonstration effect for the fully mechanised caving working face in the coal mines in China to realise the remote automatic controlling. The analysis of the working face situation is shown in Table 6-1.

Table 6-1 The analysis of the working face situation indexes

Item	September 2016	October 2016	November 2016
Mining height/m	15.7	15.8	15.7
Working face length/m	200	200	200.0
Mean footage in a month/m	188.0	226.6	267.5
Recovery ratio of the working face/%	85.13	86	88.12
Monthly production/kt	1017.37	971.7	1166.6
Retreat mining working efficiency/t · operator^{-1}	236.012	226.7	215.5
Practical working days/d	26	26	29
Influencing period of the working face accidents/h	4	2	1

7 The core technology in exploiting the mining resources in new zones

7.1 The exploration and exploiting system of the ocean ore resources [26~28]

7.1.1 Flammable ice

The natural gas hydrate is distributed in the deposit sediment in the deep ocean or the permanent frozen earth or the frozen soils in the land. As a matter of fact, it is the crystalline substances that are similar to ices, which are formed by the natural gases and water under the high pressure and the low temperature condition. Due to the fact that its external appearance is similar to ices and it can burn after it is encountered with fires, it is also called as the "Flammable ice", the "Solid gas" or the "Gassy ice".

For the resource of the flammable ice, the density is high. Furthermore, this flammable ice resource is distributed widely in the worldwide. Also, it has extremely high value of mineral resources. To the current stage, the people have already measured that in the offshore sea or the freeze soil area, the number of hydrate mine sites is more than 230. Also, a large number of enthusiastic research areas regarding the natural gas hydrates have sprung up. In China, the flammable ices are mainly distributed in the South Sea, the area of the East Sea, the Qinghai-Tibet Plateau and the frozen soil zone in the Northeast. In the continental slope of the north area of the South Sea, it is predicted that the reserves of the flammable ices are up to 19.4 billion cubic metres. As a matter of fact, the quantity like this is equivalent to 6 times of the developed reserves of the oil gases in the deep water of the South Sea. In the Xisha trough, the distribution area of the flammable ices is more than 5000 square kilometres. As for the reserves of the resources, it is up to 0.41 billion cubic metres. The molecule structure and the burning of the flammable ices are shown in Figure 7-1 and Figure 7-2.

7.1.1.1 Advantages and disadvantages of the flammable ices

The advantages of the flammable ices mainly include the following aspects:

(1) The reserves of the flammable ices are large. In the whole world, the reserves of the flammable ices are approximately 2100 trillion cubic metres. Probably, it can be used for the human beings for 1000 years.

(2) The energy of the flammable ices is high. The flammable ices with 1 cubic metres can release natural gases with 160 cubic metres.

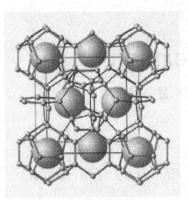

Figure 7-1 The diagram showing the molecule structure of flammable ices

Figure 7-2 The photo showing the burning of the flammable ices

(3) The burying of the flammable ices is wide. The flammable ices are distributed in the permanent frozen soils in the mainland, the deep-water environment of the interior lakes and the bottom of the sea.

(4) There is no pollution. After the flammable ices burn, there are no residue and the waste gases that will be generated.

The disadvantages of the flammable ices mainly include the following aspects:

(1) If the flammable ices are exploited not properly, it will lead to the greenhouse effect. As a matter of fact, after the mineral sources of the flammable ices are subjected to the minimum damaging, it may be possible to lead to the large quantity of the leaking of the methane gases. As a consequence, this will induce the violent greenhouse effect.

(2) On the condition that the exploiting of the flammable ices is not appropriate, it will lead to severe accidents. When the flammable ices are exploited, on the condition that the accident of the blowout occurs, it will lead to disasters such as the tsunami, the submarine landslide and the poisoning of the sea water.

Therefore, the exploiting and utilising of the flammable ices are similar to a double-edged sword. Then, it should be treated carefully.

7.1.1.2 The exploration methods of the flammable ices

A The geophysical exploration method

The seismic exploration technology is widely applied in distinguishing the storing position of the oil gases. As a matter of fact, it is also applicable for the flammable ices in the aspect of the exploitation. The essence of it is to find the Bottom Simulated Reflector (BSR). Then, the mineral resources of the flammable ices that are distributed in the large area are determined. The layer of the stable flammable ices has the relatively larger velocity of the longitudinal wave. However, in the bottom of it, the free gases that may exist, have the relatively smaller velocity of the longitudinal wave and the Poisson's ratio. Based on these specific physical parameters, the distribution range of the flammable ices can be determined. Furthermore, the reserves of the flammable ices can be calculated semi-quantitatively.

B   The geochemistry exploration method

For the geochemistry exploration method, it mainly uses that the flammable ices are extremely change with the changing of the temperature and the pressure. Additionally, in the sedimentary in the shallow area of the seafloor, particular abnormal chemistry actions are formed. Through detecting these abnormal phenomena, the position in which the flammable ices may be located can be acquired.

C   The electromagnetic technology with controllable source in the ocean

The electromagnetic technology with controllable source in the ocean means that in the seafloor or in the area that is adjacent to the seafloor, the manual stimulating is conducted and then the electromagnetic signals are received. Following this, the electric resistance of the rock strata in the seafloor is detected. Based on the character that the electric resistance of the flammable ices is relatively larger, the range of the reserves of the flammable ices, the depth of the reserves and the thickness of the reservescan be detected.

7.1.1.3   Exploiting method of the flammable ices

A   The technology with freezing drilling and hot exploitation

On May 18th, 2017, China successfully conducted trial tests to exploit the flammable ices. Specifically, the self-developed technology with freezing drilling bit and hot exploitation. This technology is developed by China. Furthermore, this technology is different from the drilling mechanism of the "coring the specimen with passive pressure maintaining and temperature maintaining", which is commonly in the world, as shown in Figure 7-3.

Figure 7-3   The "Blue Whale 1#" which is used to exploit the flammable ices

In this technology, the principle of "active coring of the specimen with reducing the temperature and freezing" is initially proposed. Furthermore, the strengthening freezing method of the shaft drilling cement-base grouts, the rapid freezing specimen coring of the hydrate in the bottom of the borehole and the thermal activation exploiting technology with high temperature pulse are invented. For the principal technical indexes, they have already been beyond the similar technologies in the overseas countries.

B The heating method

The heating method is also called as the thermal activation method. Specifically, this method means that the steam, the boiling water or the seawater with normal temperature in the surface layer are pumped into the rock strata. After that, the electromagnetism or the microwave are used to heat. The purpose of it is to make the temperature of the strata increase. Following this, the hydrate resolves. However, the disadvantage of this method is that a large amount of heat is lost. Additionally, the efficiency is low. Furthermore, collecting the gas stream is difficult, as shown in Figure 7-4.

C The exploiting method with decreasing the pressure

The exploiting method with decreasing the pressure does not need the continuous activation. As a consequence, the cost is relatively smaller. Furthermore, it is applicable for exploiting in the large area. As a matter of fact, it is especially applicable for exploiting the natural gas hydrate which is located in the underlying free gas reservoir. It is one of the most promising technologies in the traditional exploiting methods of the natural gas hydrate. Its schematic diagram is shown in Figure 7-5.

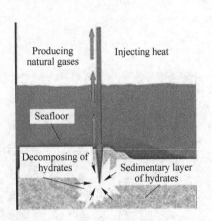
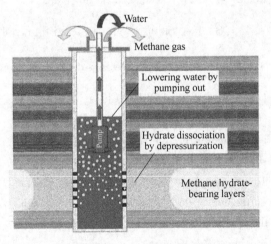

Figure7-4 The diagram showing the thermal activation exploiting method

Figure 7-5 The diagram showing the pressure decreasing exploiting method

D The exploiting method with injecting the chemical reagent

From the hole of the shaft, the chemical reagent is grouted and injected to the layer of the hydrate, such as the saline water and the methyl alcohol. This will destroy its chemical equilibrium. And this is beneficial for the decomposing of the flammable ices. However, the disadvantages of this method include the following aspects. First of all, the cost of the chemical reagents is high. Additionally, the impact of it is slow. Last but not least, it is easy to lead to pollution, as shown in Figure 7-6.

7.1.1.4 The prospect forecast of the flammable ices

The flammable ices, regarded as one of the clean resources in the future, only generate a few carbon dioxide and water after burning. The pollution of it is much smaller than the coal, the petroleum and the natural gases. However, the energy that is generated by the flammable ices is more than ten times. As a matter of fact, it is a new potential energy which has much higher heat of combustion. Furthermore, the pollution of it is much smaller and it is more cleaning. Also, the flammable ices are more convenient for using.

Once the flammable ices formally enter the industrial development direction, the relying of the human beings on the petroleum and the coal resources will decrease to a great extent. Furthermore, the human being will use the energy that is much cheaper while more efficient. The pollution of the flammable ices on the environment will be reduced for a certain extent.

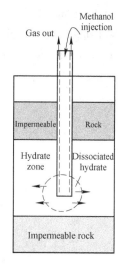

Figure 7-6 The diagram showing the method of injecting and grouting the chemical reagents

7.1.2 Manganese nodule

The manganese nodule is also called as the nodule with multiple metals, the manganese ore ball, the manganese ore mass and the manganese tumour. As a matter of fact, it is the integration of the iron and the manganese oxide. The colour of it is usually dark and the brown black. The state of the manganese nodule has many different types, such as the ball shape, the ellipse shape, the potato shape, the grape shape, the applanation shape and the slag shape. The variation of the dimension and size of the manganese nodule is also significant. Specifically, the size of the manganese nodule is ranged from several micrometres and dozens of centimetres. The maximum weight of the manganese nodule is dozens of kilograms, as shown in Figure 7-7.

Figure 7-7 Manganese nodule

7.1.2.1 Deep-sea exploiting technology of the manganese nodule

Different from the normal exploiting technology in the mainland, the exploiting technology of the manganese nodule in the deep sea includes the following aspects, namely the collecting of the mineral resources in the deep sea, the transporting system with the flexible pipe in the seafloor, the preparation of the minerals in the seafloor, the installing of the minerals in the seafloor, the power transmission and distribution, the lifting of the minerals from the deep sea to the surface of ocean, the supporting on the surface of the ocean, the positioning of the power of the mining vessel, and the monitoring and controlling in the process of the mining and the transporting. At the current stage, the developing technology of the exploiting system of the manganese nodule has already been basically mature. There are roughly the mining system with the fluid lifting (as shown in Figure 7-8), the mining system with the seafloor robots and the collecting method with the trawl.

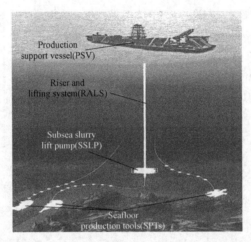

Figure 7-8 The commercial liquid lifting mining system which has already been realised by the Nautilus Minerals Limited in Canada

A The mining system with the fluid lifting

This is the core of the experimental study in the countries in the world. According to the difference in the lifting mode, this system can be divided into the lifting with the hydraulic power and the lifting with the air.

a The lifting system with the hydraulic power

This system is composed of the following four aspects, namely the mineral ore collecting instrument in the seafloor, the pump with the high pressure, the floating pipe and the mining pipe. Specifically, the mining pipe is hung below the mining vessel and the floating pipe. The function of it is transferring the manganese nodule. As for the floating pipe, it is installed in the top 15% of the mining pipe. In it, the air with high pressure is pumped and the purpose of it is supporting the water pump. The water pump with high pressure is installed in the floating pipe and the power of it is 5884kW. Through the high pressure, the lifting water flow with a high velocity of 5m/s is generated in the mining pipe. Then, the manganese nodule and the water are lifted together to the mining vessel from the seafloor. The mineral ore collecting instrument has the function of the filtering and collecting manganese nodule.

b The lifting mining system with the air

This system is basically composed of the following three parts, namely the air pump with high pressure, the mining pipe and the mineral ore collecting instrument. The air pump with high pressure is installed in the boat. When the mining is conducted, the air pump with high pressure should be initiated in the boat firstly. Then, the high-pressure air stream that is generated by the air pump moves downwards through the gas transporting pipeline, and is imported from the three sections of the mining pipe, namely the deep section of the mining pipe, the middle section of the mining pipe and the shallow section of the mining pipe. As a consequence, the composite stream with three phases of solid, air and liquid that is lifting with high velocity, is generated in the mining pipe. After that, the manganese nodule that is selected by the filtering system of the mineral ore collecting instrument, is lifted to the mining vessel. The lifting efficiency is ranged from 30% to 35%. To make that the lifting velocity of the water stream in the mining pipe arrive at 3m/s, the air compressor with the power of 4340kW must be used. The air with a volume of 225m^3 will be blown in for every second. However, the structure of this exploiting system is complicated and the fabricating cost of it is expansive. However, at the current stage, it has already been able to work at a depth of 5000m in the ocean.

B The seafloor robot mining system

The automatic mining in the seafloor is a high-level technology in mining which is relatively ideal and has the development prospect. As a matter of fact, it represents the development direction of the mining technology in the deep sea. For the automatic mining system in the seafloor, it uses the remotely controlled underwater vehicle, which can conduct the automatic mining in the seafloor. Furthermore, it can float upward automatically. Also, it can discharge the mineral ores that are collected to the semi-submersible mining platform in the ocean. Some people also call this as the underwater mining technology with robots. However, this technology requires that after each mining, it is required that the mineral ores should be delivered to the mining vessels on the sea surface. As a consequence, it has the disadvantage that the production amount is low and the reliability is poor.

C Collecting with the trawl

As a matter of fact, this is the simplest method in exploiting the manganese nodule in the seafloor. It is composed of installing a dragging net in the mining vessel. This dragging net can decline to the seafloor according to the velocity of the free falling. The loudspeaker box gage that is attached on the dragline reminds the operators when the dragging net reaches the seafloor. The dragging net canbe dragged across the seafloor. When the dragging net is fully loaded with the manganese nodules, it is retrieved. Furthermore, on the dragging net, the TV camera is installed, which is used to guide the installing operation of the dragging net.

7.1.2.2 Development prospect of the deep-sea manganese nodule mining technology

In China, the exploiting of the domestic ocean metal resources involves multiple disciplines, such as the mining, ocean, geology, atmosphere, environment, marine engineering, oceanographic engineering, mechanics, metallurgy, materials, electronic engineering and automatic control. During the process of the technical preparation, this will promote the development on the mechanical engineering, shipbuilding industry and the electronic technique in China. Therefore, this will promote these industries to enter a new stage in terms of the processing level and the equipment level. The mineral uninstalling process from the production support ship in the ocean to the mineral transporting ship is shown in Figure 7-9.

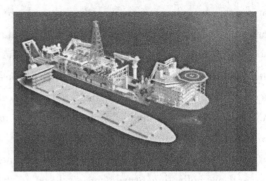

Figure 7-9 Mineral uninstalling process from the production support ship in the ocean to the mineral transporting ship

7.1.3 Petroleum

There are a large quantity of petroleum resources in the extensive ocean. According to the statistics, in the world, the ultimate reserves of the petroleum is approximately 1 trillion tons. In these resources, the exploitable reserves are 300 billion tons. Among them, the quantity of the petroleum under the ocean is 135 billion tons. At the end of the last century, the production of the petroleum under the ocean is up to 3 billion tons, which account for 50% of the whole production of the petroleum.

7.1.3.1 Domestic and international current situation in the exploiting of the deep-sea petroleum

In China, the engineering equipment level is only remained on manufacturing of parts of basement components. The equipment domestication level is low. For the matched assemblying technology of machines, it is still behindhand. It is preliminary to have the ability to design the drilling equipment in the normally deep water.

In overseas countries, the starting and developing of the ocean petroleum engineering are relatively earlier. Currently, the international ocean petroleum engineering is almost monopolised by industries in Europea, America, Japan and South Korea. Compared with that, the ocean petroleum exploiting engineering is still in the primary stage.

7.1.3.2 The exploiting technology of the petroleum in the deep sea

The petroleum exploiting technology in the sea is basically equal to the petroleum exploiting technology in the mainland. Therefore, for most technologies, the techniques used in the maintain can almost be used directly. As for the basic process, first of all, the well logging should be conducted. Specifically, the geophysical method is applied. The information regarding the original state of the rock strata that has been drilled, and the oil and gas reservoir, and that has already changed, is transmitted to the ground surface through the cable. In particular, the distribution state of the oil, the air and the water in the oil reservoir and the changing of the information. Then, based on this information, the comprehensive judging is conducted. The technical measures that should be taken are determined. After that, the shaft drilling is conducted. In the construction process, it should be guaranteed that the shaft has the minimum pollution on the oil and gas reservoir. The quality of the shaft cementing is high. As a consequence, it can bear the influence of kinds of underground operations in several tens of years of exploiting.

A Exploiting the petroleum with the natural blowing

The method of the exploiting the petroleum with the natural blowing indicates that the energy of the petroleum layer itself is fully relied to lift the raw petroleum from the bottom of the shaft to the mouth of the shaft. After that, the petroleum flows from the mouth of the shaft to the oil gathering station. At the later period, the energy in the bottom layer decreases. When the pressure that is supplied is smaller than the pressure that is consumed for lifting, the oil well stops blowing. Its schematic diagram is shown in Figure 7-10.

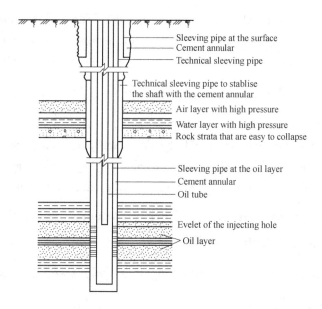

Figure 7-10 The diagram showing the structure of the naturally blowing shaft

B Exploiting the petroleum with the gas lifting

Exploiting the petroleum with the gas lifting means that the air with high pressure is artificially injected to the oil well that has already stops flowing (discontinuous flowing or the natural flowing ability is poor). The purpose of it is to decrease the gradient of the flowing pressure in the (or the composite density of the air and the hydraulic). Then, the energy of the air is used to lift the liquid, as shown in Figure 7-11.

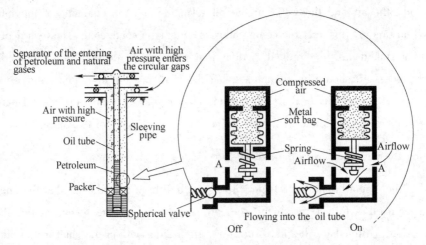

Figure 7-11 The schematic diagram showing the gas lift valve

C Exploiting the petroleum with the platform

Normally, after the fixed shaft drilling platform finishes the drilling of the cluster well, the shaft drilling equipment on the platform is replaced with the equipment that is used to exploit the petroleum. Then, the platform is changed to the fixed petroleum exploiting platform.

D The mooring system with the single point

This system is divided into two types, namely the mooring oil storage system with the single point and the mooring oil unloading equipment with the single point. It is applicable for the oil field in which the production small. Additionally, it is applicable for the oil field that is relatively far away from the coast or the oil field in which the evaluative trial production is conducted.

7.2 The distribution and the exploiting expectation of the mineral resources in the polar region [29, 30]

7.2.1 The mineral resources in the polar region is various and abundant

7.2.1.1 The mineral resources in the area of the North Pole

A The petroleum resources in the North Pole

The natural resources of the North Pole region are extremely abundant. As a matter of fact, they include the mineral resources that are non-renewable, the chemical energies, the biological resources that are renewable and the unchanging resources such as the hydraulic power and wind power. According to the generalised definition of the resources, the military resources, the scientific resources, the humanistic resources and the tourism resources should be included.

According to the conservative estimation, the potential reserves of the petroleum that is exploitable in this region is ranged from 100 billion buckets to 200 billion buckets. As for the natural gases, they are ranged from 50 trillion cubic metres to 80 trillion cubic metres. It can be found that when the resources of the petroleum and gases in the other areas in the world are going be exhausted, the North Pole will become the last energy base for the human beings.

B The coal resources in the North Pole

In the north of the Alaska, the resources of the coal are abundant. As a matter of fact, it is one of the areas that have not been exploited. The geologists estimate that 9% of the total coal resources in the world are buried in this place, which are approximately 400 billion tons. As a matter of fact, this is even superior than the "coal capital"-Datong in Shanxi Province in China, which is famous not only domestically but also overseas. In the west of the North Pole, the theoretical reserves of the coal resources are 0.3 billion tons. Among the coal basins in the north of the Alaska, it is the coal basin with the highest quality. Furthermore, this coal basin can be exploited with the normal open-pit excavating and mining technology. The reserves of the coal resources in Siberia are larger than the Datong of China and the Alaska of the North America. Someone estimates that it is more than 700 billion tons or even larger. Moreover, it is even more than half of the total coal reserves of the world.

In the North Pole, the coal resources are not only abundant. As a matter of fact, the quality of the coal masses is superior. The coal masses in the western area experience the ancient geological forming process for 0.1 billion years. It is the bituminous coal with high-volatility. The average calorific value is more than 12000 Joule per Kilo. The sulphur is low, which is about 0.1% to 0.3%. Also, the ash is low, which is about 10%. Moreover, the temperature is lower, which includes 5% water. The coal in the North Pole is almost the cleanest in the world, which has extremely high steam and coking quality. It can be directly used for energy and the industrial raw materials.

C The iron ore and other mineral resources in the North Pole

In the North Pole, the other mineral resources are also abundant. The world-level iron mine in the Prince Charles Mountains, is well-known throughout the world. Besides the iron ore, the North Pole also has a large amount of other mineral resources. In Norilsk, the largest composite mine basement of the Nickel Plutonium in the world is one of them. In Alaska, it is estimated that the Blue Dog mine site which is located in the north of the Kurtz, possesses 85 million tons of ores. Among them, the zinc occupies for 17% and the lead occupies for 5%. As for the silver, there are 75 grams for each ton. It has already become the world-level large mine site with the value of 11.1 billion US dollars (the value in 1983).

In the gold mine area in Alaska, from 1880 to 1943, approximately 108.5 tons of gold are produced. It is estimated that there are still 13.2 tons of gold that need to be exploited. The Chichagov Mine that is adjacent to the Sitka, produced 24.8 tons of gold in the past. There are still 9.3 tons of gold to be exploited. As for the exploiting of the precious metal ores, it is different in both sides of the Bering strait. Additionally, the silver mine of the Green Creek, is the largest potential silver mine in the full area of the United States of America. After it is exploited since 1988, the production ability is processing 1000 tons of minerals for one day. It is estimated that this mine site can be exploited for 10 years to 30 years.

Besides the above-mentioned mineral resources, this place also has the radioactive elements, such as the uranium and the plutonium. They are called as the strategic mineral resources. For example, in the Prince of the Wales Island, the salt included mine has 0.285 million tons of plutonium ores.

7.2.1.2 The mineral resources in the Antarctic region

The Antarctica, is located in the surrounding area of the South Pole. Most of it is located in the south of the Antarctic circle (66°64′S). It is the mainland that is covered by the ice and snow. Around it, the islands are scattered all over like stars in the sky or men on a chessboard. The area of the Antarctica, includes the mainland of the South Pole and the islands. Specifically, it is approximately 14 million square kilometres, accounting for 10% of the area of the mainland of the world. It is the fifth largest mainland in the world.

A The petroleum and the natural gases in the Antarctica

The petroleum and the natural gases in the Antarctica are mainly distributed in the continental shelf of the Antarctica. As a matter of fact, they are one of the resources that most catches human being's eyes. It is generally believed that the Ross Sea and the Ross Ice Shelf, the Weddell Sea and the Filsine Ice Shelf Area, the Prydz Bay and the Lambert Garden Area are the main exploration area in which the potential of the resources including the petroleum and natural gases is largest. The continental shelf of the Antarctic Peninsula, the Bellinsgauzen Sea and the Amundsen Sea Shelfare also the favourable exploration area in which the displaying of the petroleum and the gases are favourable. The New Zealand and the sea area such as the Gaussberg-Kerr Galen, the Klose and the seafloor highland of Falkland, are the prospective areas to seek the resources such as the petroleum and gases. The reserves of the petroleum in the Antarctic region is approximately ranged from 50 billion buckets to 100 billion buckets. As for the reserves of the natural gases, they are ranged from approximately 3000 billion cubic metres to 5000 billion cubic metres. The schematic diagram of the ice layer and bed rock in South Pole is shown in Figure 7-12.

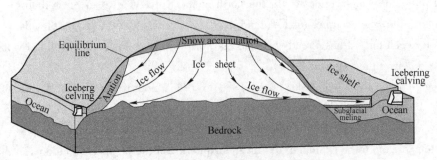

Figure 7-12 The ice layer and the bed rock in South Pole

B The coal resources in the Antarctic Pole

The Permian coal seam of the Antarctic mainland is mainly distributed below the ice cover of the Antarctica. The reserves are approximately 500 billion tons. The thickness of the coal seam is approximately 2m. The largest thickness is 10.7m. The shape of the coal seam is usually the lentoid. Along the strike direction, the extending of the coal seam is usually less than 3m. The scientific investigation data show that the Permian coal seam of the Antarctic mainland is mainly distributed below the ice cover of the Antarctica. The reserves are approximately 500 billion tons. To the current stage, the coal basins that have already been found are mainly distributed in the section of the transantarctic mountains of the Antarctica along the Ross Coast. Additionally, in the Ellsworth mountainous area which is located in the West of the Antarctica, there is also appearing of the coal basin. The coal basin that is located transantarctic mountains of the Antarctica, is probably the largest coal basin in the world.

At present, in the transantarctic mountains of the Antarctica, the coal that is exposed on the ground surface have already been found. Its coal seam thickness is approximately 6m to 8m. As a matter of fact, it is one of the biggest coal basins in the world. In the surrounding area of the Beaver Lake that is in the north of the Prince Charles Mountains, the thickness of the coal seams that are exposed is ranged from 2.5m to 3.5m. Furthermore, it is the superior power coal.

C The iron mines and other mineral resources in the South Pole

The iron ore is one of the mineral resources which the South Pole is in rich of. In the Antarctic mainland, they are mainly distributed in the southeast of the Antarctica. In 1966, according to the exploration of the scientists, in the north of the Mountain Rucker which is located in the south of the Prince Charles Mountains, the strip shaped rock strata with rich magnetite deposit is found. Specifically, the thickness of it is 70m and the average iron content is 32.1%. In 1977, below the ice cover that is in the west of the Mountain Rucker, the region in which the magnetism is abnormal is found. Specifically, the length of it is ranged from 120km to 180km. The width of it is ranged from 5km to 10km. In the south of the Westfall Hill, the zone in which the magnetism is abnormal is found. Specifically, the length of it is 120km and the width of it is 24km. Through analysing, it is believed that the magnetic anomaly of this zone is resulted by the magnetite. And this is the Iron Mine of the Prince Charles Mountains, which is called as the "Iron Mountain" in the Antarctica mainland. The average grade of these iron ores is ranged from 32% to 58%. As a matter of fact, it is the ore deposit with rich irons, which have industrial exploiting values. It is preliminarily estimated that the reserves of it can be utilised by the world for 200 years. Furthermore, it is one of the largest iron mines in the world.

The Antarctica is in rich of abundant mineral resources. Besides the iron ore and the coal resources, at the current stage, the number of other mineral resources that have already been discovered is more than 220. These include the copper, the lead, the zinc, the aluminium, the gold, the silver, the graphite, the diamond and the petroleum. Additionally, there are also rare mineral deposits which have important strategic values, such as the thorium, the plutonium and the uranium.

7.2.2 Exploiting the resources in the polar region

7.2.2.1 Exploration of the resources in the polar region

The coring and the drilling in the ice layer in the deep of the polar region includes the drilling of the overlying accumulated snow layer, drilling of the ice layer and the drilling of the rock strata below the ice.

A Drilling in the overlying accumulated snow layer

Above the ice cover in the polar region, it is the accumulated snow layer. First of all, the accumulated snow layer that is above the ice layer should be drilled through. Then, the drilling can be conducted on the ice layer that has the truly significance. Due the difference of the climate and the geographical condition, the thickness of the accumulated snow layer in different regions is significantly different. In some areas, the thickness of the overlying accumulated snow layer is up to more than 100 metres. Due to the fact that the accumulated snow layer has high permeability, drilling into this kind of rock strata is pretty complicated. Generally, when drilling into the ice snow layer, to prevent the leaking of the shaft drilling liquid from the bottom of the annular pipes, the isolation of the annular pipes should be conducted and the particularly sealed casting shoes should be used. Additionally, on the condition that the depth of the accumulated snow layer is not large, the shaft drilling liquid is not needed to balance the pressure of the rock strata. Then, the coring and drilling in the dry holes can be conducted.

B Drilling in the ice layer

Ice is a particular rock stone which is a non-linear rheology medium. Even though the stress is not large, it still cannot yield the ice. Furthermore, this will lead to the creeping of the ice itself. Then, decreasing of the hole diameter will be induced, which may further lead to the collapsing of the borehole. Due to the fact that approximately 80% of the jamming of the drilling bit in the coring and drilling of the deep ice layer is resulted by the creeping of the ice layer, the drilling in the ice layer needs the shaft drilling liquid whose density is appropriate to balance the ground pressure of the rock strata. This is to guarantee the long-term stability of the borehole wall.

C The drilling in the rock strata below the ice

At the current stage, there are two modes, namely the drilling in the ice cover and the drilling in the rock strata below the glacier. In one mode, the electric machinery with the armour cable type is used to conduct the drilling while in other mode, as shown in Figure 7-13. For other type, the commercial slewing drilling machine is used to conduct the drilling. Due to the fact that the slewing drilling operation is quite not stable, and the weight and power consumption of the commercial slewing drilling machine are relatively large, this method is therefore not appropriate for the drilling in the ice layer.

At the current stage, the most effective method to drill into the junction between the ice layer and the bed rocks is the drilling technology with the electric machinery with the armour cable. In the drilling of the ice layer, the drilling bit with hardness alloy cutting is used, to drill into the ice layer and the soft ice layer. When the harness grade of the rock strata below the ice arrives at IX ~ X, the coring and drilling bit with the diamond attached is used, to drilling into the ices and the rock strata in the basement.

7.2.2.2 Exploiting the petroleum in the polar region

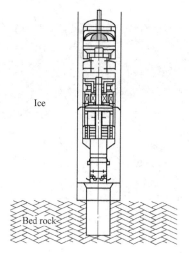

Figure 7-13 The electric machinery with the armour cable type

A The common methods used in the exploiting of the petroleum

The main methods used in the exploiting of the petroleum include pressing the boiling water or the water vapour with high temperature. Even burning of part of the petroleum under the ground will be used. Sometimes, the nitrogen is pressed in. Also, the carbon dioxide is pressed to decrease the viscosity of the petroleum. Similar to this, the light petrol will be pressed to decrease the viscosity of the petroleum. The aqueous solution of the organic matter is pressed, to decompose of the oil from the rock strata. Additionally, the aqueous solution of the substance (such as the detergent) which can improve the surface tension (the interface tensiometer) between the oil and the water is pressed, to decompose of the oil from the rock strata. Besides this, according to the different situations, two measures or more than two measures should be adopted to conduct the composite exploiting.

B The equipment that is commonly used in the petroleum exploiting

In the well drilling of the petroleum, the drilling tool is used to crush the rock stones. And the drilling is conducted towards the underground. The borehole with the stipulated depth is drilled out, for the petroleum exploiting machine or the gas exploiting machine to extract the petroleum or the natural gases. The petroleum drilling machine that is commonly used is mainly composed of eight segments, including the power machine, the driving motor, the working machine and the auxiliary equipment, as shown in Figure 7-14.

The pumping unit in the petroleum exploiting is usually called as the "Kowtow machine". Specifically, the method of increasing the pressure is used to extract the petroleum from the shaft, as shown in Figure 7-15. When the pumping unit is in the upstroke, the pipeline shrinks elastically and moves upwards. This induces the mechanical petroleum exploiting instrument move upwards. Then, it impacts the sliding sleeve and generates vibration. At the same time, the check valve to the positive direction is shut down. The piston whose diameter changes is used to seal the petroleum.

Figure 7-14　The petroleum drilling machine　　　Figure 7-15　Petroleum pumping unit

When the pumping unit in the petroleum exploiting is in the downstroke, the pipeline elastically elongates and moves downwards. This will lead to the mechanical pumping unit moving downwards. Furthermore, it will impact the sliding sleeve, generating vibration. At the same time, the reverse check valve will be partially shut down. The piston whose diameter changes is still used to seal the oil sleeve. Therefore, in the bottom area of the reverse check value, high pressure zone is formed. This movement generates a reverse impaction force on the channel of the petroleum in the rock strata.

C　The well drilling platform in the area that is adjacent to the North Pole in Alaska

The United States of America is one of the pioneer countries that are focused on the exploiting of the petroleum and utilising of the petroleum. As a matter of fact, its petroleum exploiting technology is leading in the world. Similar to most of the other petroleum producing counties, the petroleum exploiting technology of the United States of America also experiences three development stages, including the exploiting with the natural blowing, the exploiting with the mechanical equipment and the tertiary petroleum exploiting technology. Among these three methods, the exploiting with the mechanical equipment is the most commonly used method. The United States of America is one of the pioneer countries that conduct the experiments of the tertiary petroleum exploiting technology and the countries that have the largest application scale. Specifically, the tertiary petroleum exploiting technologymainly includes the chemical petroleum driving technology and the petroleum driving technology with CO_2. It should be noted that the petroleum driving technology with CO_2 is one of the most widely applied technologies in the United States of America to improve the recycling ratio. Therefore, America is one of the countries that have the largest application scale in the petroleum driving with CO_2 and the countries that improving the recycling efficiency. The well drilling platform and the petroleum exploiting are shown in Figure 7-16~Figure 7-18.

Since 1986, the petroleum driving technology with the chemical method is showing the decreasing trend. In particular, the petroleum driving with the surface-active agent is almost ceased. However, using the polymer to adjust still has a large amount of development prospect, especially the deep

Figure 7-16　In 1996, the well drilling platform that is adjacent to the delta of Colville in the North Pole of the Alaska in USA

Figure 7-17　The well drilling platform near the North Pole of Alaska in USA

Figure 7-18　Petroleum exploiting in Alaska

adjusting. As a matter of fact, it is not the sole measure to increase the production. Under the certain condition, it can replace the petroleum driving with the polymer. Or it can be combined with the polymer, making the petroleum driving with the polymer have larger efficiency.

Compared with the petroleum driving with the polymer, the quantity of the chemical agents that are needed is relatively smaller. Furthermore, the investment recovery is rapid. Among them, the new deep adjusting system (Colloidal Dispersion Gel or CDG) is commonly paid much attention in recent years. This gel is one kind of the semi-fluid state. As a matter of fact, it can flow gradually in the rock strata. Due to the fact that the concentration of the polymer is pretty low, the cost will be decreased largely. At the same time, due to the fact that the viscosity of the material is quite small, this will not make the pressure that is injected increase apparently. As a consequence, the quantity that enters the strata with low permeability will relatively decrease. Furthermore, due to the fact that they are connected gradually, they can enter the deeper area of the rock strata, largely improving the adjusting effect.

7.2.3 Exploiting expectation of the resources in the polar region

In the 21st century, we have already entered the era of the polar region. In recent years, especially the initial ten years in the 21st century, due to the warming up of the climate, the thawing of the sea ices in the polar regionmakes this area become more apparent and significant in the strategic position in the international politics. As a matter of fact, the polar region is only the investigation and the studying interest point of the scientists. At the current stage, it is pushed to the leading edge of the geopolitics. Because of the abundant resources, the polar region is gradually becoming the focus of the governments of all countries and the academic world. Under the game of the superpower countries, the problem of the rights and interests in the polar region and the problem of the solving in the polar region are becoming one of the hotspots and sensitive issues in the internal politics.

As for the South Pole, due to the extremely severe natural environment, comprehensively considering the feasibility of the exploiting and economics of the feasibility, resource exploiting will not be conducted at the current stage. In the South Pole, China has already established the Great Wall Station and the Zhongshan Station. Additionally, the China's research institution in the pole, the state archives centre in the pole and the China's data information management system in the poly are established. The ice-breaker with the level of 20 thousand level is possessed, and this ice-breaker is used for transportation and the scientific investigation. 12 expedition teams for the Antarctica have already been dispatched. Successively, there are 5 ocean research vessels, which have already been the South Pole for 11 times. The travelling is more than 250 thousand sea miles, transporting more than 2000 scientific workers and the engineering technical staff to the South Pole to conduct the investigation. The initial comprehensive scientific investigation in the Southern Ocean of China is accomplished. Additionally, the South Pole scientific investigation of "Seventh Five-Year" plan and the "Eighth Five-Year" plan is conducted. The years of observation on the meteorology, the physics of the earth, aeronomy and the radio transmission are conducted. Additionally, the investigation and the research on multiple disciplines in terms of the glacier, the geology, the landform, the environment, the surveying and mapping, the mining, the biology, the human medicine and the oceanography are conducted. Based on the work, a large quantity of valuable materials, specimens and data are collected.

7.3 Exploiting characters of the highland mining industry, and the exploiting technology and equipment [31, 32]

7.3.1 The mineral resources in the highland and the Alpine region

7.3.1.1 Geological situation of the Qinghai-Tibet Plateau

The highland and the Alpine region indicate that the altitude is high. The temperature is low. In the winter, it is cold while in the summer it is cool. Additionally, the frost-free season is short. The air is thin and the transparency is high. The solar radiation is strong and the sunlight is long. There are more days without raining. The temperature difference between the daytime and the night is large. The precipitation rainfall capacity is relatively small. However, the temperature is low and the evaporation capacity is weak. There are a quantity of snow mountains. Furthermore, the glaciers are located everywhere. The sunshine is strong and the solar energy is abundant, as shown in Figure 7-19.

Figure 7-19 The Loess Plateau

7.3.1.2 The profile of the mineral resources in the Qinghai-Tibet Plateau

The mineral resources are abundant. The Qinghai-Tibet Plateau has a vast territory. It is located in the Tethys-the Himalaya mineral resource area. The geological structure is complicated and the effect of the mineralisation is violent. As a matter of fact, it is one of the best areas with the potential to explore the minerals. In this area, the mineral resources are pretty abundant. The types of the minerals are relatively complete. In particular, the mineral resources of the nonferrous metal are extremely abundant. On August 12, 2007, the China Geological Survey announced in Beijing that the geological survey in the Qinghai-Tibet Plateau which has been conducted for 7 years and invested with 0.34 billion Chinese Yuan, have acquired significant findings. In the Qinghai-Tibet Plateau, there are more than 600 ore deposits, mineral occurrence and the mineralised spots, which have been newly found. These include the significant mineral resources which are extremely in shortage in China, such as the copper, the iron, the lead and the zinc. Furthermore, it is predicted that the prospective resources of the copper in the Qinghai-Tibet Plateau are ranged from 30 million tons to 40 million tons. Furthermore, the prospective resources of the lead and the zinc are 40 million tons. Additionally, the prospective resources of the iron are several billion tons. The specialists that are in charge of this geological surveying point out that to the current stage, more than 90% of the iron mines that have already been discovered in China are low grade ores. This time, there are totally 3 iron-enriched mines that are discovered in the Qinghai-Tibet Plateau. All of them have reached the large scale of. Additionally, in the Qinghai-Tibet Plateau, the new alternative energy, namely the oil shale. This kind of resource can be exploited and transformed to the petroleum.

The potential in the exploiting of resources is large. In the Qinghai-Tibet Plateau, the potential of the mineral resources is enormous. As a matter of fact, it has the geological conditions in exploring the scare minerals of the country and the large-scale ore deposits. It can become the important mineral resource basement of China. In the Qinghai-Tibet, the copper, the iron, the lead, the gold, the salt lake minerals (the boron and the lithium), the terrestrial heat and the alkaline mineral water are the superior minerals. Additionally, there are mineral resources that have potential advantages, such as the antimony, the molybdenum, the rare metals (the rubidium, the caesium and the strontium), the sylvite, the limestone which is used for cements and the non-metallic ores that are used for construction, such as the granite. For the iron deposits that is in shortage in China, the reserves of the discovered ore deposits are 5.196 million tons. The recoverable reserves are 3.645 million tons. The grade of the mineral ore deposits is high. Most of them are in the metallurgy level. In the copper deposit zone of the Yulong porphyry, the reserves of the discovered mineral resources are 9.0605 billion tons. The are 4 large-scale ore deposits and 1 medium-scale ore deposit.

7.3.2 The technic technologies of the open-pit mining in the highland mining

In Tibet, the average altitude is more than 4000 metres. In the winter, it is extremely cold. The overall natural environment is relative severe. This leads to the great challenge for the operation of the mine sites. Based on the distribution characters of the solid mineral resources in the highland, and the severe environment in the highland area, at the current stage, the open-pit mining is more widely used in the exploiting of the solid mineral resources in the highland area.

7.3.2.1 The technic technology of the open-pit mining in the highland

In the production process of the open-pit mining, there are mainly four production techniques:

(1) Borehole drilling and blasting procedure. The blasting or the mechanical methods are used to loosen and crush the mineral rocks on the stairs. This is appropriate for the equipment for

excavation.

(2) The operation of mining and tunnelling. The excavating equipment is used to load the loose mineral rocks on the stairs to the transporting equipment. As a matter of fact, this is the core of the open-pit mining.

(3) Transporting. It means that the transporting equipment is used to deliver the mineral rocks to the scenes and the reserving heap. As for the waste rocks, they are delivered to the waste dump.

(4) The operation of discharging. It means the unloading operation of the mineral rocks and the discharging operation of the rocks.

Generally, the open-pit mining is composed of four principal links, namely the borehole drilling, the mining, the transporting and the discharging. Or it means the borehole drilling and blasting, the excavating and the mining, the transporting and the land discharging. Sometimes, the drainage should be conducted. After the mining activities, it is needed to consider the environmental governance. There is no specific stipulation on the equipment that is needed to accomplish these procedures. At the current stage, the commonly used methods include the down-hole drill, the anger drill, kinds of explosive materials, the electric shovel, the hydraulic excavator, the lifting van shovel, the truck, the bulldozer and the land leveller. The new technologies and new techniques are encouraged by the state. However, this must be checked by the state.

A The technique of the borehole drilling

The technique of the borehole drilling can be classified according to the energy and the utilising form of the drilling. Specifically, it can be divided into the following two aspects, namely the borehole drilling with the thermal crushing and the borehole drilling with the mechanical crushing. The borehole drilling equipment generally include the impact drilling with the steel wire rope, the fire drill and the drill jumbo. Additionally, the new drilling machines for the borehole drilling with the ultrasonic wave, the borehole drilling with chemical materials and the borehole drilling with the high-pressure water are being developed. The pinch roller drilling machine (KY-250B) is shown in Figure 7-20.

Figure 7-20 The pinch roller drilling machine (KY-250B)

B The blasting technology

The blasting technology is part that is necessary in the production process of the mine site. Specifically, the expanding and application of the blasting technology in the micro-borehole and the millisecond blasting technology, successfully addresses the problems of the crushing blackness and the vibration decreasing in blasting in the rock strata that are difficult to be blasted in the open-pit mining and excavating process. At the current stage, kinds of new explosive materials and blasting instruments are continuously improving the efficiency of the blasting, the quality of the blasting, the safety of the operation. Then, the information about the millisecond blasting technology will be given. At the current stage, for the blasting technology, a large number of open-pit mine sites in China are using the millisecond blasting technology, in which the plastic detonating tube and the non-electro-explosive system is used. Although it is restricted by the operation condition, generally, the small-scale blasting operation is conducted. Nevertheless, it can still satisfy the stripping quantity of approximately 2d to 3d for each electric shovel. Following this, the information about the pre-splitting blasting technology will be given. In the places that are adjacent to the slopes, the application of the pre-splitting blasting technology will decrease the influence of the vibration effect on the slope to a large extent. The flatness of the stair surface can be guaranteed. As for the twice crushing, it means that the jack drill is used to conduct the borehole drilling. Then, the explosive materials are adopted to conduct the blasting. To increase the producing efficiency of the large mineral blocks, the method of the concentrated twice crushing can be used. The electric detonator can be used to conduct the instantaneous detonating.

C The exploiting techniques

a The exploiting technology in the steep slope

The characters of the exploiting technology in the steep slope are that in the initial stage, the quantity of the stripping is relatively small. Additionally, the construction period of the infrastructure is short. Furthermore, the amount of the engineering is small. Last but not least, the time that the slope is exposed is short. Therefore, for the exploiting of the open-pit mine sites in the current stage of China, it is extremely appropriate.

b The exploiting technology on high stairs

Due to the fact that the research on this kind of technology is relatively late in China, this technology is not widely applied in the open-pit mine sites. The maximum height of the stairs is 14mm to 15mm. However, as the mining and excavating equipment in the open-pit mining is gradually developing towards the large-scale direction, the research on the mining technology with high stairs is also continuously increasing. In recent years, the increasing of the large-scale mining and excavating equipment with the bucket capacity larger than $10m^3$, provides the advantageous technical guaranteeing for the mining technology with high stairs.

c The mining technology with the cemented backfilling

The characters of the cemented backfilling are that it can significantly improve the recycling ratio of the mineral resources. Furthermore, it can effectively decrease the destruction rate of the ground surface. At the same time, it can realise the multiple mining of the open-pit mine sites. Therefore, it is always the preferentially selected method in the mining industry. However, its disadvantages are also extremely apparent. The developing of the efficiency of this technology is restricted by the production condition of the ore deposits. With the developing of the technologies such as the backfilling in the high sub-layer, backfilling in the sub-sections and the subsequent backfilling in the deep boreholes in the stage, the mining technology with the cemented backfilling is still continuously developing.

D The dumping technology

When the open-pit mining activities are being conducted, whether the dumping efficiency is high or low directly influences the production scale. Additionally, at the same time, it will influence the economic benefits. For the normal condition, the dumping method is selected according to the developing form and the transporting form. However, the parameter that influences the utilising time of it is the stability of the structure of the refuse dump. The dumping form of vehicles-dumping plough, has the advantages of flexible manoeuvring, strong ability in climbing the slope, being able to adapt to abundant landforms, low cost and being convenient for maintaining. Because of these advantages, it is widely selected by the mine sites.

E Waterproof and drainage technology

Due to the fact that the producing operation of the open-pit mining is all conducted in the open-pit condition, to guarantee that in the flood season, the mine site can still work properly, the corresponding waterproof and drainage technology is extremely important. The waterproof and drainage technology can guarantee the safety of the operation in the flood season. At the same time, it can prevent the flowing of the ground surface water. Furthermore, the quantity of the discharged water and the water content in the mineral rocks can be decreased. This technology mainly uses the measures such as setting the intercepting drain, flood bank and rechannelling of the rivers, to intercept and dredge.

7.3.2.2 The open-pit mining equipment in the highland

From the perspective of the equipment level, the open-pit mining equipment in China can be basically classified as three types. Small-scale: the equipment is simplex. Mismatching of the equipment occurs. For some procedures, the manual labour is still adopted at the current stage. Medium-scale mine site and some large-scale mine sites: the degree of mechanisation is relatively higher. However, for some equipment, they fall behind. And the efficiency is not high. Small number of large-scale mine sites: the imported or the domestically manufactured modern equipment are used, such as the geared drill, the electric-wheel truck, the excavator with large bucket capacity, the bulldozer with large power. They constitute the set of the operating line and the efficiency is relatively high.

In China, the set of the typical equipment for the open-pit mine site has two different levels. In the first level, the electric shovel with the volume of $4m^3$ is the main part. Additionally, the down-the-hole drill of 200mm, the geared drill of 250mm, the tilting cart of 60 tons and 108 tons, the electric locomotive of 100 tons and 150 tons, the self-discharging vehicle of 20 tons, 27 tons and 32 tons, and the bulldozer of 132.3kW and 161.7kW are matched. In the second level, the electric shovel with the volume of $10m^3$ is the main part. Additionally, the geared drill of 310mm, the self-discharging vehicle of 60 tons, the electric-wheel truck of 100 tons and 154 tons, the bulldozer of 235.2kW, the loading machine of 5m^3, the explosive loading truck of 15 tons and the road roller of 40 tons are matched, as shown in Figure 7-21.

(a) (b)

(c) (d)

(e) (f)

(g)   (h)

Figure 7-21　The equipment used in the open-pit mining

(a) The electric shovel; (b) The dumping plough; (c) The open-pit down-the-hill drill with high wind pressure KQZ120Y; (d) The geared drill KY-250B; (e) The Volvo excavator EC360; (f) The loading machine with the wheel type ZL30E; (g) The mine site used truck; (h) The self-discharging vehicle

7.3.2.3　The development direction of the open-pit mining techniques in the highland

At the current stage, in China, the open-pit mining technical technology is becoming mature. However, the equipment is relatively behindhand. Therefore, the principal development direction of the open-pit mining in China is increasing the research intensity on the open-pit mining equipment. At the current stage, the development direction of the open-pit mining equipment is the large scale, the automation and the intelligence.

A　The vehicle-mounted monitoring system

The function of the vehicle-mounted equipment is monitoring the operating state of the equipment. This provides the corresponding information for the maintaining of the equipment. For the large-scale open-pit mining equipment, adding the warning and monitoring functions, can realise the real time monitoring of the operating state of the equipment. When the equipment shows problems, the notification can be sent to the maintenance personnel at the first time to main and process. Then, the consumption on the equipment can be decreased and the maintaining cost on the equipment can be reduced.

B　Displacement monitoring of the high slope

The displacement monitoring of the high slope is realised with the positioning system of the GPS satellites. Specifically, through the wireless communication network, it can continuously send the displacement data of the slope to the computers that are in the offices in the mining site. Furthermore, storing and computing analysis can be conducted on the data, which is beneficial for accurately understanding the stability situation of the kinds of slopes in the mine. Furthermore, it can provide the function of early warning.

C The environmental protection on the equipment is increased

Currently, the energy is short and the environmental problems are becoming severe in China. Based on these issues, the equipment designing personnel pay an increasing number of attentions to the energy consumption of the mining equipment and the pollution of the mining equipment. In the prospective designing of the open-pit mining equipment, the long service lift, the low energy consumption and thematerials that have low environment pollution and high performance, are all the designing principles that should be followed.

References

[1] Qian Minggao. Efforts to be made to achieve the transition of Chinese coal industry from quantity to quality [J]. China Coal, 2017, 43(7): 5-9.

[2] Zhu Wancheng, Guan Kai, Yan Baoxu, et al. Development tendency of mining engineering and future demands for the knowledge structures [J]. Education Teaching Forum, 2018(30): 6-10.

[3] Liu Jie. Construction and practices of the Tongmei Datang Tashan Coal Mine Co. Ltd. with high production, high efficiency and high safety. Special report on the safety acceptance of the Tashan Coal Mine[R]. 2008.

[4] Shangwan Coal Mine in the Shendong Coal Mine Group. Special report on the production situation of the fully mechanised working face with extremely large mining height of 8.8m[R]. 2019.

[5] International Exchange and Cooperation Center, State Administration of Work Safety, United States (U.S.) National Security Council. Zhong meikuangshanan quanhe zuoxiangmu peixunjiaocai [M]. 2005.

[6] Marcus A Wiley, Gregory Abramov. Borehole mining: getting more versatile [J]. Mining Engineering, 2004, 56 (1):36-40.

[7] Zhang Fengwei, Zhen Xuan, Chen Chuanxi. Development status and tendency of world open-pit coal mine [J]. China Coal, 2014, 40(11): 113-116.

[8] He Fulian, Chen Dongdong, Xie Shengrong. The kDL effect on the first fracture of main roof with elastic foundation boundary [J]. Chinese Journal of Rock Mechanics and Engineering, 2017, 36(6): 1384-1399.

[9] Li Chunlin. Special academic report on the rock bolt reinforcement and the strengthening of roadways [D]. University of Science and Technology Beijing, 2006.

[10] Wu Jingke. Study on roof disaster-causing mechanism and control technology of gob-side entry retaining in deep coal mine[D]. China University of Mining and Technology, 2017.

[11] Kang Hongpu. Controlling technologies and application of the roadway surrounding rock masses in deep coal mines[R]. In the special report on the academic forum of the deep roadway surrounding rock mass controlling and the highly efficient recovery engineering of coal resources. 2005.

[12] Chen Yanguang, Lu Shiliang. Strata control around coal mine roadways in China [M]. Xuzhou: China University of Mining and Technology Press, 1994.

[13] Huai Bei Kuang Ye (Ji Tuan) Co. Ltd., China University of Mining and Technology, Shijiazhuang Ao Xun Kuang Yong She Bei Co. Ltd. Research report on the instability mechanism of the deep rock roadway floor in coal mines and the integrated technology of locking and grouting[R]. 2010.

[14] Huai Bei Kuang Ye Co. Ltd., Anhui University of Science and Technology. Rock bolt reinforcement in the loosen mudstones with high stress and extremely loose thick coal seam, and the mutual strengthening between supports[R]. 2012.

[15] Xuzhou Kuang Wu Ji Tuan Co. Ltd., Ping Liang Xin An Mei Ye Co. Ltd. Research report on the soft rock roadway reinforcement technology and application in the Xinan deep coal mine[R]. 2013.

[16] SDIC Xinji Group. Research report on the exploring and application of the reinforcement technology for roadways with a thousand metres in depth in the Kouzidong Mine[R]. 2013.

[17] Feng Xiating, Chen Bingrui, Zhang Chuanqing, et al. Mechanism warning and dynamic control of rockburst development processes [M]. Beijing: Science Press, 2013.

[18] Yang Chunhe. Academic Special Topic Report. Progress report on the research on the energy storage underground chamber disaster mechanism and the preventing theory[R]. 2011.

[19] Miao Xiexing. New progress on the backfill coal mining theory and technology. Academic Special Topic Report[R]. 2011.

[20] Zhou Huaqiang. Paste backfill to recover the remained coal pillars by the strip mining. Academic Special Topic Report[R]. 2011.

[21] He Fulian, Kang Rong, Li Hongbin, et al. Mining techniques and technology of waste gangue replacement in soft-broken coal-rock region under high stress [J]. Journal of China Coal Society, 2011, 36(1): 1-6.

[22] Li Shugang. Theory and technology of the combined mining of coal and gas based on the elliptic paraboloid zone in the mining induced fractures. Academic Special Topic Report[R]. 2011.

[23] Gu Dazhao. Technology and engineering of the underground reservoir in coal mines. Special report in the scientific forum regarding the deep mine roadway surrounding rock mass controlling and the high recovery engineering of coal[R]. 2015.

[24] Datong Coal Mine Group, Tiandi Co. Ltd., Tongxin Coal Mine Co. Ltd. Core technology and demonstration engineering on the intelligent controlling of the fully mechanised caving working face with more than ten million tons[R]. 2017.

[25] Yu Bin, Xu Gang, Huang Zhizeng, et al. Theory and its key technology framework of intelligentized fully-mechanized caving mining in extremely thick coal seam [J]. Journal of China Coal Society, 2019, 44(1): 42-53.

[26] Wang Zaiyi. Evolution report on advanced marine technologies [M]. Qingdao: China Ocean University Press, 2017.

[27] Tian Lengzhu. Hai yang shiyoukai caigong cheng [M]. Dongying: China University of Petroleum Press, 2015.

[28] Xu Junliang, Ren Hong, Wang Zhifeng, et al. Primary research on key technology for deep water gas hydrates drilling and sampling [M]. Beijing: Petroleum Industry Press, 2014.

[29] Xiao Jun, Yuan Zhijun. Eluosibeiji quyude jinshukuang chanziyuan[J]. World Nonferrous Metals, 2009(8): 72-73.

[30] Liu Aihua, Liu Xibing, Zhao Guoyan. Teshukuang chanziyuan kaicaifang fayujishu [M]. Changsha: Central South University Press, 2009.

[31] Qiao Junwei, Li Congcong, Fan Qi, et al. Characteristics of coal resources and their geological background at Northern Qinghai Tibet Plateau [J]. Journal of China Coal Society, 2016(2): 294-302.

[32] Long Tao. Research and practice of mining development and ecological harmony in Tibetan Plateau [J]. Mining R&D, 2014, 34(1): 113-115.